Lab Manual to Accompany

Fifth Edition

AGRISCIENCE

FUNDAMENTALS AND APPLICATIONS

Delmar is proud
to support FFA activities

Join us on the web at

agriculture.delmar.com

Lab Manual to Accompany

Fifth Edition

AGRISCIENCE

FUNDAMENTALS AND APPLICATIONS

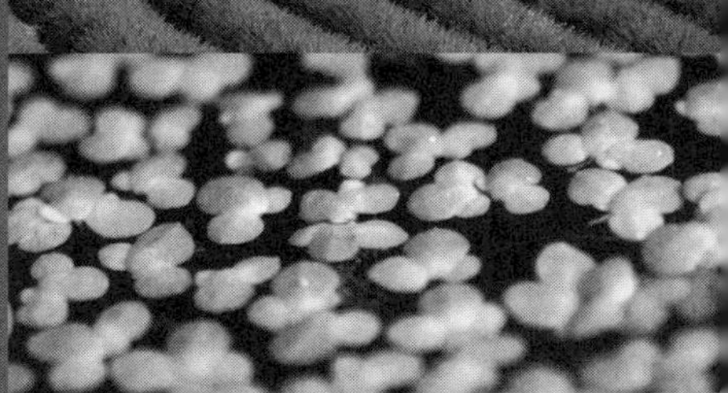

L. DeVere Burton
Renee J. Peagh
Walter York

DELMAR
CENGAGE Learning™

Australia • Brazil • Japan • Korea • Mexico • Singapore • Spain • United Kingdom • United States

DELMAR
CENGAGE Learning™

Lab Manual to Accompany
Agriscience: Fundamentals and Applications, 5th Edition
L. DeVere Burton, Renee J. Peugh, Walter York

Vice President, Career and Professional Editorial: Dave Garza

Director of Learning Solutions: Matthew Kane

Acquisitions Editor: Benjamin Penner

Managing Editor: Marah Bellegarde

Product Manager: Christina Gifford

Editorial Assistant: Scott Royael

Vice President, Career and Professional Marketing: Jennifer McAvey

Marketing Director: Debbie Yarnell

Marketing Manager: Erin Brennan

Marketing Coordinator: Jonathan Sheehan

Production Director: Carolyn Miller

Production Manager: Andrew Crouth

Content Project Manager: Katie Wachtl

Senior Art Director: David Arsenault

Technology Project Manager: Tom Smith

Production Technology Analyst: Thomas Stover

For product information and technology assistance, contact us at
Professional & Career Group Customer Support, 1-800-648-7450

For permission to use material from this text or product, submit all requests online at **cengage.com/permissions**
Further permissions questions can be emailed to
permissionrequest@cengage.com

Library of Congress Control Number: 2008942493

ISBN-13: 978-1-4354-1968-1

ISBN-10: 1-4354-1968-5

Delmar
5 Maxwell Drive
Clifton Park, NY 12065-2919
USA

Cengage Learning products are represented in Canada by Nelson Education, Ltd.

To learn more about Delmar, visit **www.cengage.com/delmar**

Visit our corporate website at **www.cengage.com**

Notice to the Reader
Publisher does not warrant or guarantee any of the products described herein or perform any independent analysis in connection with any of the product information contained herein. Publisher does not assume, and expressly disclaims, any obligation to obtain and include information other than that provided to it by the manufacturer. The reader is expressly warned to consider and adopt all safety precautions that might be indicated by the activities described herein and to avoid all potential hazards. By following the instructions contained herein, the reader willingly assumes all risks in connection with such instructions. The publisher makes no representations or warranties of any kind, including but not limited to, the warranties of fitness for particular purpose or merchantability, nor are any such representations implied with respect to the material set forth herein, and the publisher takes no responsibility with respect to such material. The publisher shall not be liable for any special, consequential, or exemplary damages resulting, in whole or part, from the readers' use of, or reliance upon, this material.

Printed in the United States of America
3 4 5 6 7 14 13 12 11 10

Contents

Preface ... vii

Section 1 Agriscience in the Information Age

Exercise 1 There's a Method to the Madness—The Scientific Method. 2
Exercise 2 Getting Up Close and Personal—The Microscope ... 5
Exercise 3 It's Alive!—The Cell ... 10
Exercise 4 Don't Drink the Water: An Insight into Water Safety ... 15
Exercise 5 How Degrading—Maintaining a Healthy Environment ... 17

Section 2 You and the New Millennium

Exercise 6 Get a Job—Your Career in Agriscience ... 24
Exercise 7 Experience Required—Your SAEP ... 35
Exercise 8 Show Me the Way— Leadership in Agriscience ... 40

Section 3 Natural Resources Management

Exercise 9 Watt's the Big Deal—Creating Electricity Out of Thin Air (An exercise adapted from a National Renewable Energy Laboratory Booklet) ... 48
Exercise 10 Getting Down and Dirty—Preventing Soil Erosion ... 50
Exercise 11 Talking Dirty—Soil Fertility ... 54
Exercise 12 Hold the Soil Please—Growing Plants without Soil ... 58
Exercise 13 There's Something Fishy Here—Fish Anatomy ... 61

Section 4 Integrated Pest Management

Exercise 14 Are You a Competitor? Weed Identification lab ... 72
Exercise 15 Don't Bug Me—Insect Anatomy ... 74
Exercise 16 Stuck on You—External Parasites ... 78

Section 5 Plant Sciences

Exercise 17 Putting Down Some Roots—Root Anatomy ... 82
Exercise 18 It Stems from—Stem Anatomy ... 86
Exercise 19 Turn Over a New Leaf—Leaf Anatomy ... 90
Exercise 20 He Loves Me, He Loves Me Not—Flower Anatomy ... 93
Exercise 21 Under the Knife—Flower Dissection ... 97
Exercise 22 Free Gas Made Here—Photosynthesis ... 100
Exercise 23 Transpiration—Nature's Big Gulp ... 103
Exercise 24 Reap What You Sow—Seed Anatomy ... 105

Exercise 25 Come Out, Come Out Whereven You Are—Germination . 108
Exercise 26 Let's See What You're Made of—DNA Extraction 112

Section 6 Crop Science

Exercise 27 We Are Here, We Are Here, We Are Here—Soil Organisms 116
Exercise 28 A Good Start—Growing and Transplanting Vegetable Seedlings 120
Exercise 29 Try That Oon For Size—Grafting Fruit Trees 124
Exercise 30 From Fries to Fuel—Making Biodiesel 127
Exercise 31 Deep Thoughts—Exploring Seed Planting Depths 129
Exercise 32 Does This Smell Good?—Storing Silage 132
Exercise 33 Stop Buggin' Me—Insect Collection 135

Section 7 Ornamental Use of Plants

Exercise 34 Keep 'em in the Dark—Plants and Light 140
Exercise 35 Chip Off the Old Block—Plant Cloning 143

Section 8 Animal Sciences

Exercise 36 You Eat Like a Pig—Simple Animal Digestion 148
Exercise 37 It Takes Four—Ruminant Digestion 152
Exercise 38 Must Have Been Something I Ate—Internal Parasites 157
Exercise 39 It's All in the Genes—Understanding Genetics 162
Exercise 40 This Will Hurt Only for a Minute—Injection Procedures.. 166
Exercise 41 We Are Doing What?—Fecal Examination for Parasites .. 169
Exercise 42 Hot Wings Anyone?—Animal Anatomy and Physiology .. 172

Section 9 Food Science and Technology

Exercise 43 Break It on Down?—Food Nutrients.................... 178
Exercise 44 What's for Dinner?—Exploring the Food Guide Pyramid . 182
Exercise 45 Is This Still Good?—Food Preservation................. 184
Exercise 46 How Cheesy—Dairy Product Production................ 186
Exercise 47 Does This Milk Make Me Look Fat?—Homogenization ... 188
Exercise 48 Do You Have a Sweet Tooth?—Fruit Sugars 190

Section 10 Communications and Management in Agriscience

Exercise 49 Where Did the Money Go?—Understanding Diminishing Returns... 194
Exercise 50 I Owe You—Interest and Credit 199
Exercise 51 In the Money—Writing a Business Plan................. 203
Exercise 52 It's a Balancing Act—Balance Sheets 205
Exercise 53 Rich Man, Poor Man–Profit–Loss Statements 209
Exercise 54 Money, Money, Money—Interest 212
Exercise 55 Can't We All Just Get Along—Cooperative Math Simulations 215

Preface

This laboratory manual was developed to accompany *Agriscience: Fundamentals & Applications,* Fifth edition. The topics and exercises throughout were designed to meet the science-based curriculum needs of the Agriscience 1 course in secondary schools. Although written with the organization of *Agriscience: Fundamentals & Applications,* Fifth edition, in mind, this can be used as a stand-alone laboratory manual.

Equipment and Supplies List

The equipment and supplies listed here were carefully selected to be used with the exercises in the *Lab Manual to Accompany Agriscience: Fundamentals & Applications,* Fifth edition. All items listed here are available from Carolina Biological Supply Company, 2700 York Road, Burlington, NC 27215. You may also place your order or request a catalog by calling toll free 1-800-334-5551.

	Carolina Biological Supply Catalog Number
Exercise 2	
Compound microscope	59–1244
Beginner's Slide Set, which includes a letter "e," newsprint, colored threads, cork, cornstarch, salt crystals, dust, cotton fibers, nylon fibers, silk fibers, wool fibers, and volcanic ash slides	
Exercise 3	
Compound microscope	59–1244
Animal cell slide	23575
Plant cell slide	B1066
Chromosomes—Mitosis slide	B551
Mitochondria slide	B554
Golgi apparatus slide	H335
Onion epidermis slide	B600A
Glass microscope slides	63–2950
Coverslips	63–2960
Medicine droppers	736898

	Carolina Biological Supply Catalog Number
Exercise 4	
Pond water	16–3380
Medicine droppers	
Glass microscope slides and coverslips	
Microscopes with 30X and 100X power	
Exercise 5	
Paper strips	
Dried plant materials such as leaves	
Scissors	64–4770
Spray bottle	
Water	
Three plastic cups	
Sand	
Topsoil	
Plastic	
Biodegradable plastic	
Cotton fibers	
Synthetic fibers	
Magnifying glass	
Gloves	

	Carolina Biological Supply Catalog Number
Exercise 6	
Thumbtack	
Sheet of posterboard—18 inches × 24 inches	
Scissors	
Glue	
Occupational Outlook Handbook	
Dictionary of Occupational Titles	
Exercise 7	
Official FFA Manual	
SAEP Record Book	
Exercise 8	
Official FFA Manual	
Parliamentary Guide for FFA or Robert's Rules of Order	
Exercise 9	
100 cm of copper wire	
1 bar magnet	
1 electrical meter	
1 cardboard tube	
Duct tape	
Fan or turn table	
Exercise 10	
Six plant trays (or shoe boxes cut to 5 inches deep)	
Plastic sheeting to line trays	
Six sprinkling cans	
Six beakers	
Trowel	
Scissors	
Stopwatches	
Soil	
Sod (grass-covered soil from a lawn)	
Water	
Supports to form inclines for trays	

	Carolina Biological Supply Catalog Number
Exercise 11	
Seven identical 6-inch flower pots with drainage outlets at the bottom	
Pot chip or screen	
Vermiculite or perlite	
Mixing bin or box	
Pea and tomato seeds	
Nitrogen source—NH_4NO_3	
Phosphorus source—superphosphate	
Potassium source—KCI	
Calcium source—$CaCO_3$	
Materials for labeling pots	
Laboratory balance or scale	
Ruler or yardstick	
Exercise 12	
Hydroponics Experiment Kit	
Personal Safety Set	19–9998
Exercise 13	
Textbook	
Dissection Lab Perch Kit	
One pair of scissors	
One scalpel	
One sharp probe	
Magnifying glass	
Personal Safety Set	19–9998
Exercise 14	
A large amount of Newspaper	
Two pieces of tempered hard board, peg board or plywood (~12" × 18" rectangles)	
Cardboard cut into ~12" × 18" rectangles	
Tie downs or bungee cords for strapping the weed portfolio tightly	
Large pieces of construction paper for mounting the specimens	

	Carolina Biological Supply Catalog Number
Exercise 14 *(cont'd.)*	
Plastic bags for collecting (old grocery bags work well and keep the specimens moist)	
Shovel	
Adhesive for mounting the specimens	
Blank adhesive labels	
20 weeds collected near your home	
Weed Identification manuals	
Small zip style bags	
Exercise 15	
Dissection Manual Grasshopper Kit	
One pair of fine dissection scissors	
One sharp probe	
Dissecting microscope	
Personal Safety Set	
Exercise 16	
Compound microscope (used for most lice and fleas)	
Dissecting scope (sufficient for larger lice and engorged ticks)	
Insecticide powder	
Animal comb or any inexpensive plastic hair comb	
Smooth dark material (paper or plastic)	
70 percent ethyl alcohol or isopropyl (rubbing) alcohol	
Glass jar	
Water or oil (baby oil)	
Microscope slides and cover slips	
Dissecting needle	
Personal Safety Set	
Exercise 17	
Monocot root slide	
Dicot root slide	

	Carolina Biological Supply Catalog Number
Compound microscope	
Set of colored pencils	64–4285
Exercise 18	
Monocot stem slide	
Dicot stem slide	
Compound microscope	
Set of colored pencils	
Exercise 19	
Leaf section slide	
Compound microscope	
Set of colored pencils	
Exercise 20	
Taraxacum young bud slide	
Taraxacum older bud slide	
Mature anther slide	
Pollen germination slide	
Compound microscope	
Set of colored pencils	
Exercise 21	
A complete flower	
Dissecting probe (teasing needles)	
Scissors	
Scalpel (optional)	
Tweezers/forceps (optional)	
Dissecting scope or zoom lens 18 to 108 mm	
Exercise 22	
Calomba plant or other water plant	
Manometer	
Water	
Large beakers	
High-intensity light source	
Light-tight box	
Exercise 23	
2 geranium leaves	
2 small jars (baby food or Mason jars, or 2 beakers)	

Carolina Biological Supply Catalog Number

Exercise 23 *(cont'd.)*
Cardboard (10 cm square with a hole about the size of a hole-punch in center)
Clay or petroleum jelly

Exercise 24
Seed of a plant slide
Compound microscope
Several fresh bean seeds
Magnifying lens
Scalpel

Exercise 25
Water
Aluminum foil
Beaker
Hot plate
Freezer
Seed-Gro Kit

Exercise 26
1 plastic quart-size bag
1 strawberry
10 mL DNA extraction buffer
Filter paper
Ice cold 90 percent ethanol or Isopropyl alcohol
1 Test tube or champagne flute
Funnel (optional)
Inoculation loop or coffee stir straw
Microscope
Slides and Coversheets

Exercise 27
What's in the Soil II—Microorganism Kit
Compound microscope—300X or better
Four 400-mL beakers
Test tube rack
Two test tubes
Bunsen burner

Carolina Biological Supply Catalog Number

Handful of green leaves
Soil samples
Water
Personal Safety Set

Exercise 28
Tomato seeds
Seed-starting media
Flats
Spray bottle with water
Trowel or spoon
Plastic wrap
Large containers (10 oz. or larger)
Plant-growing media
Liquid fertilizer
Ruler
Toothpicks

Exercise 29
Knife
Grafting wax
Grafting tape
Apple or other fruit stock—¼- to ½-inch diameter
Scions—¼- to ½-inch diameter

Exercise 30
1 liter *new* vegetable oil
4 grams substance A
250 mL substance B
2 glass containers with lids (glass juice bottle)
1 funnel
4 qt pot
Warming plate
Thermometer
Blender (Do not use with food again)
Turkey baster (Do not use with food again)

Exercise 31
Six glass jars—pint size
Soil

Carolina Biological Supply Catalog Number

Exercise 31 *(cont'd.)*
Water
Corn seeds

Exercise 32
Freshly cut chopped grass
Six 1-quart jars with tight-fitting lids
Labels for the jars
Pen
Nails
Hammer

Exercise 33
Hand lens 10–30X power, or a dissecting scope
Sweep net (can be purchased from any biological educational catalog)
Stapler
Paper bags (large grocery store size)
Glass jars with screw-top lids
Ethanol or isopropyl alcohol (rubbing alcohol)
Use of a freezer (optional)
Glass dish or Petri dish

Exercise 34
Six pots of pea seedlings
Four light-tight boxes, each with a hole in one side
One completely light-tight box
Light source

Exercise 35
Plant II Cloning: Propagation of African Violets
Kit
African violet plant
Soap
Bunsen burner
Scalpel
Water—Sterile
Four to five large beakers

Carolina Biological Supply Catalog Number

African violet potting soil
Ethyl alcohol
Hot plate
Laundry bleach
Ten to twelve small plastic pots
Personal Safety Set

Exercise 36

Item	Catalog Number
Textbook	
Rubber gloves	
Scalpel	
Dissecting microscope	
Personal Safety Set	19–9998
Preserved digestive tract of a pig or a fetal pig	

Exercise 37

Item	Catalog Number
Textbook	
Digestive tract of a pig or fetal pig	P1905C
Rubber gloves	19–9817
Scalpel	62–5920
Four jars	
Dissecting microscope	59–1822
Compound microscope	
Rumen bacteria slide	
Personal Safety Set	19–9998
Preserved digestive tract of a cow or sheep	

Exercise 38

Item	Catalog Number
Sheep Liver Fluke Microscope Slide Set	
Sheep Tapeworm Microscope Slide Set	
Pig Roundworm Microscope Slide Set	
Compound microscope	59–1244

Exercise 39
Calculator

Exercise 40
Two or three 10-mL syringes
½-inch needles

	Carolina Biological Supply Catalog Number
Exercise 40 *(cont'd.)*	
1-inch needles	
Three containers of water	
Food dye—red, green, blue	
Three or four grapefruit	
Ruler	
Exercise 41	
Compound microscope	
Plastic bag (Ziploc® type)	
Jelly bean size of fresh feces of a cat, dog, sheep, cattle, hog, or horse	
Microscope slide and coverslip	
Small piece of newspaper	
Toothpick	
Tap water	
Eyedropper	
Gloves and eye protection	
Exercise 42	
One raw chicken wing for each student. (Chicken wings can be purchased at the local supermarket in the inexpensive family pack.)	
Scissors	
Paper plate	
Tweezers or forceps	
Protective gloves	
Exercise 43	
Twenty test tubes	
Test tube rack	
Five eyedroppers	
Hot plate	
Sudan IV	
Biuret reagent	
Potassium iodide—Iodine solution	
Benedict's solution	
Milk	
Cream or butter	
Syrup	
Canned whole corn	
Distilled water	
Exercise 44	
Variety of foods from each food group	
Kitchen measuring tools	
Scale in ounces	
Exercise 45	
Bread	
Swiss cheese	
Hot plate	70–1015
20-mL test tubes	19–8915A
250-mL beaker	72–1209
Distilled water	
Test tube holder	70–2900
Sulfuric acid solution	89–3411
Funnel	73–4012
Filter paper	71–2742
Spatula	70–2742
Ethyl alcohol	86–1281
Exercise 46	
12-inch piece of cheesecloth	71–2690
Hot plate	70–1015
Plastic cups—16 ounces each	
Cooking thermometer	74–5384
Milk	
Salt	
Rennin or rennilase	20–2375
Exercise 47	
Compound microscope with an oil immersion lens	
Microscope slide and coverslip	
Raw milk from a dairy or a health food store that carries raw milk	
Homogenized milk	

	Carolina Biological Supply Catalog Number
Exercise 48	
Refractometer	
A variety of fruits	
Kitchen knife	
Blender (optional)	
Exercise 49	
Calculator	
Exerscise 50	
Calculator	
Exercise 51	
Computer	
Calculator	91–2442
Exercise 52	
Calculator	91–2442
Exercise 53	
Calculator	91–2442
Exercise 54	
Calculator	91–2442

SECTION ONE

Agriscience in the Information Age

EXERCISE 1: There's a Method to the Madness— The Scientific Method

Materials Needed

- ✓ Textbook
- ✓ Other materials deemed appropriate by students

Purpose

"The scientific method" is best described as the steps scientists and students follow when trying to explain the world around them. In this exercise, you will learn and use the scientific method.

Procedure

1. Study the steps of the scientific method below. These steps will help you throughout this laboratory manual.

   Step 1. Through observation, identify the problem. Ask the question: "What, exactly, do I want to find out? Limit your topic to a single research objective.

   Step 2. Review the literature. Read up on and become familiar with the topic.

   Step 3. Form a hypothesis. Develop a hypothesis or statement to be proven or disproven.

   Step 4. Prepare a project proposal. Outline how the project should be done. Include timelines, facilities, equipment required, costs, and a description of how you will do the project.

   Step 5. Design the experiment. Develop a plan for carrying out the project.

   Step 6. Collect the data. Conduct the experiment. Record what you measure or observe.

   Step 7. Draw conclusions. Summarize the results. Make calculations. Determine if the information collected allows you to accept or reject the hypothesis.

   Step 8. Prepare a written report. This report should include your hypothesis, a brief project description, results, and the conclusions. This report will provide you and others with a permanent record of your research.

Observations

2. Using the scientific method in agriscience

   Step 1. Identify the problem. The following problem has been identified for you. Complete the remaining steps of the scientific method.

   How does the color of a material affect the heat and light absorption or reflection?

Step 2. Review the literature. In the space provided, list any references that you used for review. Include the name of the references and other pertinent information.

____

____

____

____

Step 3. Form a hypothesis. Write your hypothesis in the space provided.

____

____

____

____

Step 4. Prepare a project proposal. Outline your project in the space provided. Include timelines, facilities, equipment required, costs, and a description of how you will do the project.

____

____

____

____

____

____

Step 5. Design the experiment. Develop your plan to test your hypothesis in the space provided.

____

____

____

____

Step 6. Collect the data. Record any data in the space provided. Design your own data format.

____

____

____

____

Step 7. Draw conclusions. Summarize your results in the space provided. Confirm or reject your hypothesis.

Step 8. Prepare a written report. Using a computer and word processing program, prepare a written report not to exceed two pages explaining your project and the results.

Conclusions

1. How did the scientific method help you to prove or disprove your hypothesis?

2. Which step in the scientific method do you consider to be the most important and why?

3. Which step in the scientific method do you consider to be the least important and why?

EXERCISE 2: Getting Up Close and Personal—The Microscope

Materials Needed

- ✓ Compound Microscope
- ✓ Beginner's Slide Set, which includes a letter "e," newsprint, colored threads, cork, cornstarch, salt crystals, dust, cotton fibers, nylon fibers, silk fibers, wool fibers, and volcanic ash slides

Purpose

The microscope is an amazing tool that allows you to see objects that are too small to be seen with the naked eye. In this exercise, you will be introduced to the compound microscope. You will learn microscope anatomy, how to use one, and finally practice using it.

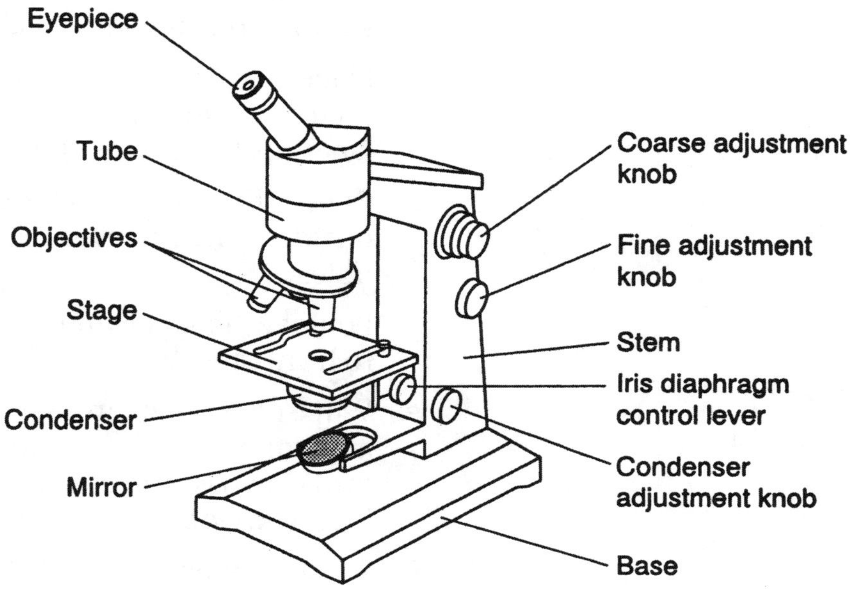

Procedure

1. Study the preceding diagram of a microscope. Learn the names and functions of the parts.

   a. **Eyepiece**—magnifying lens at the top of the microscope

   b. **Tube**—connects eyepiece to the objectives

   c. **Objectives**—lens that magnifies the object to be viewed

   d. **Stage**—platform that holds the slide for viewing

   e. **Condenser**—concentrates the light source onto the object being observed

   f. **Light source**—provides light to view the slide on the stage; it may be a mirror or a lamp.

   g. **Coarse-adjustment knob**—knob used for initial focusing

   h. **Fine-adjustment knob**—knob used for final focusing

   i. **Stem**—connects the base to the lens assembly

j. **Iris-diaphragm control lever**—controls the amount of light to the slide

k. **Condenser-adjustment knob**—adjusts light brilliance through the condenser

l. **Base**—platform to which entire microscope is attached; sits on table or desk

2. Prepare the microscope and view the slide.

a. If your microscope has a mirror, position the mirror so that light is reflected through the hole on the stage. If your microscope has a light, turn it on.

b. Place the letter "e" microscope slide on the stage and secure it with the stage clips. Make sure the coverslip faces up. Center the slide over the stage hole.

c. Rotate the low-power objective so that it locks into position directly over the specimen and over the stage hole.

d. Look at the stage and use the coarse-adjustment knob to bring the objective down within the coverslip.

e. While observing through the eyepiece, rotate the coarse-adjustment knob counterclockwise to bring the image into view. Use the fine-adjustment knob for sharp focusing. ***Remember:*** *Always focus upward.*

f. Practice opening and closing the iris diaphragm by moving the lever on the side of the condenser clockwise to provide more light and counterclockwise to provide less light. More light is needed for higher-power objectives. Less light is required for transparent objects. Record your observations in Observation 2.

g. While looking through the eyepiece, observe the letter "e" slide while moving the slide forward, backward, left, and right. Describe what happens to the "e" in Observation 3.

h. Carefully turn the objective to the high-power lens. Look through the eyepiece at the letter "e" slide. If it is not in focus, adjust the fine-adjustment knob only. (Never use the coarse-adjustment knob to adjust the high-power objective lens because this lens is very close to the slide and could break it.)

i. Observe each of the other slides found in the beginner's slide set. Draw a picture of or describe what you see for each slide in Observation 4 that follows.

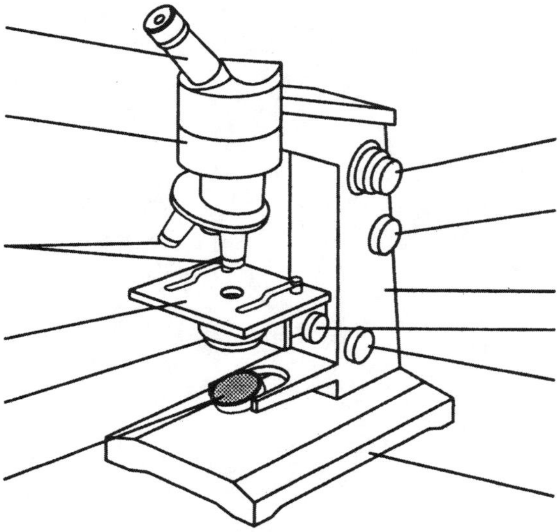

Observations

1. Label the parts of the microscope.

2. Describe the effects of opening and closing the diaphragm.

____

____

____

____

____

3. Describe the appearance of the letter "e" under the microscope and how it moves as the slide moves.

____

____

____

____

____

4. Draw a picture of or describe in detail each of the following as it appears under the microscope.

   a. newspaper print

   b. colored threads

c. cork

d. cornstarch

e. salt crystals

f. dust

g. cotton fibers

h. nylon fibers

i. silk fibers

j. wool fibers

k. volcanic ash

Conclusions

1. How does the image you see through the microscope appear when compared with the non-magnified object seen directly?

2. Explain how the movement of the slide on the stage and the movement as viewed through the microscope differ.

3. In what areas of agricultural research would the microscope be useful? List at least five.

EXERCISE 3: It's Alive!—The Cell

Materials Needed

- ✓ Compound microscope
- ✓ Animal cell slide
- ✓ Plant cell slide
- ✓ Chromosomes—Mitosis slide
- ✓ Mitochondria slide
- ✓ Golgi apparatus slide
- ✓ Onion epidermis slide
- ✓ Glass microscope slides
- Coverslips
- Medicine droppers
- Cell Reference Materials

Purpose

To study and compare the structures of plant and animal cells and to identify the common parts of cells.

Animal Cell

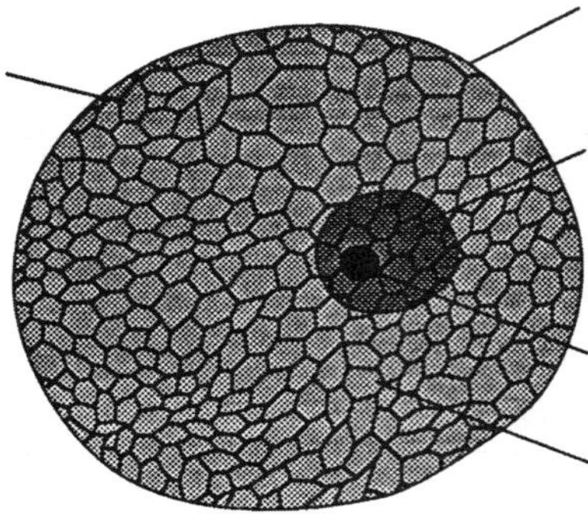

Plant Cell

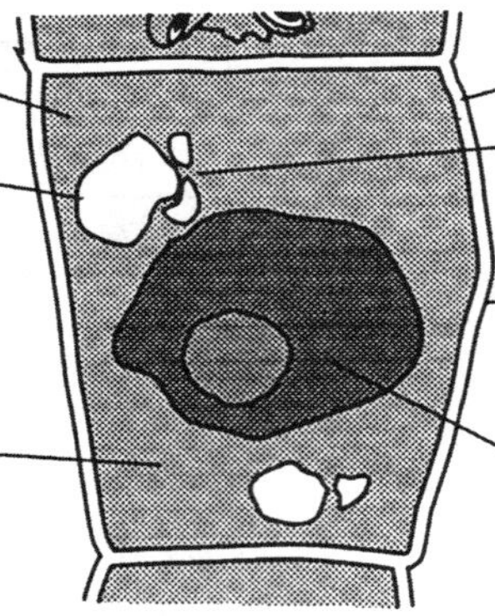

Procedure

1. Viewing animal cells

   a. Place the general animal cell slide on the stage of the microscope and view it under low power.

   b. Switch to high power. In Observation 1 that follows, identify these parts: cell membrane, cytoplasm, nucleus, nucleolus, and nuclear envelope.

   c. Place the chromosome slide on the stage of the microscope and view it under low power.

d. Switch to high power. In Observation 2, identify the chromosomes in the drawing.

e. Place the mitochondria slide on the stage of the microscope and view it under low power.

f. Switch to high power. In Observation 3, identify the mitochondria in the drawing.

g. Place the Golgi apparatus slide on the stage of the microscope and view it under low power.

h. Switch to high power. In Observation 4, identify the Golgi apparatus in the drawing.

i. Use the general animal cell picture to help you complete the statements in Observation 5.

2. Viewing plant cells

a. Place the general plant cell slide on the stage of the microscope and view it under low power.

b. Switch to high power. In Observation 6, identify the following parts: cell wall, plasma membrane, cytoplasm, nucleus, central vacuole, chloroplasts, and plasmodesmata.

c. Use the general plant cell picture to help you complete the statements in Observation 7.

3. Viewing onion cells

a. Place the onion epidermis slide on the stage of the microscope and view it under low power.

b. Switch to high power. In Observation 8, draw an onion cell and identify the following parts: cell wall, plasma membrane, cytoplasm, nucleus, and vacuoles.

Observations

1. Using the following drawing, identify the following:

cell membrane
nucleus
nuclear membrane
cytoplasm (cytosol)
rough endoplasmic reticulum
lipiddroplets

centrosomes
chromosomes
mitochondria
Golgi Body (Apparatus)
ribosome cluster

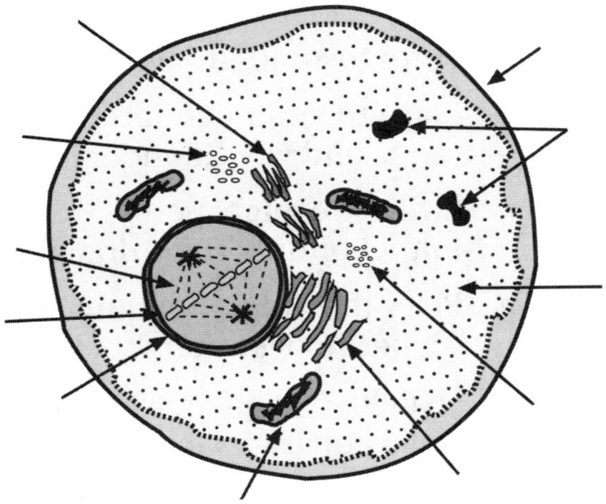

2. Using the Internet and any other cell reference materials provided by your instructor, fill in the blanks to make the following statements true:

   a. The ______________ ______________ is the outermost structure of an animal cell.
   b. The fluid medium in which much of the intermediary metabolism of the cell takes place is the ______________.
   c. The large organelle that contains the cell's inherited information and serves to control and coordinate all metabolism is the ______________.
   d. The ______________ separates the nuclear contents from the rest of the cell.
   e. The ______________ are tangled strands that extend throughout the nucleus and contain genes.
   f. Cellular organelles in which the chemical energy in food molecules is converted to a form useful to cellular metabolism are called ______________.
   g. The ______________ plays a part in the storage or secretion of metabolic products.
   h. The site of ribosomal RNA synthesis and ribosome assembly is the ______________.

3. Identify the parts of the general plant cell on the following diagram.

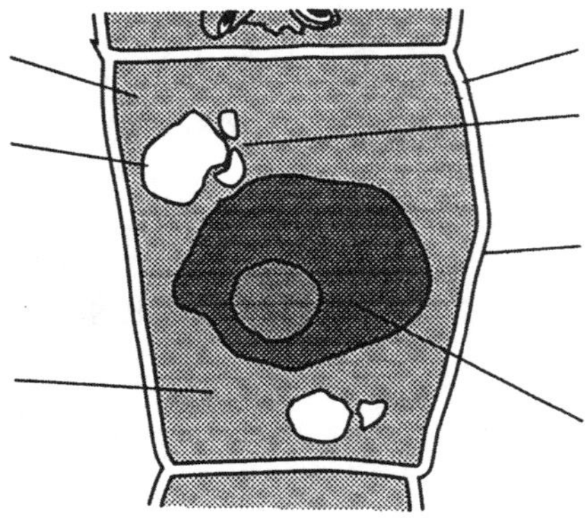

4. Fill in the blanks to make the following statements true.

   a. The ______________ ______________ is the outermost structure of a plant cell.

   b. The membrane-enclosed sac that occupies up to 90 percent of the cell volume is the ______________.

   c. The large organelle that contains the cell's inherited information and serves to control and coordinate all metabolism is the ______________.

   d. The ______________ consists of a viscous fluid matrix that contains water, many organic and inorganic substances, and a variety of tiny, specialized, functional structures.

   e. The ______________ forms the outer boundary of a cell and maintains the chemical composition of the cell by regulating the passage of substances in and out of the cell.

   f. ______________ are organelles that often contain lamella; chloroplasts are an example.

5. Draw an onion cell and label its parts.

Conclusions

1. What parts are different between animal and plant cells?

____

____

____

2. What are the differences between the general plant cell and the onion cell?

3. What are some other plant cells that would be different from the general plant cell?

4. What parts of the general animal and plant cells are similar?

5. What differences can be observed in the general shapes of the animal and plant cells?

EXERCISE 4: Don't Drink the Water: An Insight into Water Safety

Materials Needed

- ✓ Pond water
- ✓ Medicine droppers
- ✓ Glass microscope slides and coverslips
- ✓ Microscopes with 30X and 100X power
- ✓ Internet access or a pond water organism field reference book

Purpose

In many parts of the world people do not have access to clean potable (drinkable) water. In more developed countries, water purification systems are a vital part of daily living. After completing this exercise, you will have a greater appreciation for these systems. You will also be able to observe and identify some small organisms that live in still water.

Procedure

1. Each student will collect a small jar of water from a local pond. Allow for the water to sit in the class room near a window for a few days prior to the lab for incubation. Note: river and stream water should not be used; they will not produce favorable results.

2. Place one drop of water onto the slide. Place your coverslip onto the slide by holding the coverslip at a 45° angle to the water then gently drop it onto the slide.

3. Examine the slide under the microscope at 30X. When you have found an organism, adjust the microscope to 100X.

Using the Internet or reference books, try to identify your organism. To find a good identification key on the Internet, use keywords "pond water key." Guides may also be found under the "Images" browsers using key words "pond water organisms."

Observations

1. Draw a picture of three organisms you found.

2. How does each organism move?

3. Identify any organelles you may be able to see, and label them on your original drawing.

Conclusions

1. What are the names of the organisms you found?

2. Do you think these organisms are single celled or multicellular? Why?

3. Compare and contrast the organisms you found.

4. Do you think the organisms you found may be harmful to humans?

5. On a scale from 1 to 10, how important are water purification systems? Why?

EXERCISE 5: How Degrading—Maintaining a Healthy Environment

Materials Needed

- ✓ Land Fills and Environmental Kit
- ✓ Paper strips
- ✓ Dried plant materials such as leaves
- ✓ Scissors
- ✓ Spray bottle
- ✓ Water
- ✓ Three plastic cups
- ✓ Sand
- ✓ Topsoil
- ✓ Plastic
- ✓ Biodegradable plastic
- ✓ Cotton fibers
- ✓ Synthetic fibers
- ✓ Magnifying glass
- ✓ Gloves

Purpose

To determine the solid-waste materials that are biodegradable and to explore some of the factors that influence the rate of degradation.

Procedure

Note: Students may work individually or in teams of three.

1. Preparing samples

a. Cut the paper into ½-by-2-inch strips. Cut the two kinds of plastic into ½-by-2-inch strips.

b. Cut the synthetic and natural fibers into 2-inch lengths.

c. Cut the dried plant materials into pieces that are about 2 inches long and no wider than ½ of an inch.

d. Separate the materials into three piles. Each pile should have equal parts of all of the materials being tested.

2. Preparing the landfill

a. Label three plastic cups as follows:

Cup #1—Sand

Cup #2—Dry topsoil

Cup #3—Moist topsoil

b. Fill cup #1 with sand, up to 1 inch from the top, packing the sand tightly.

c. Fill cups #2 and #3 with topsoil, up to 1? inch from the top, packing the soil lightly.

d. Water the topsoil in cup #3 using a spray bottle. Water until soil is moist but not excessively so that there is water at the bottom of the cup. The soil should be uniformly moist and not have any residual water pooled at the bottom of the cup.

3. Adding materials to the landfill

a. Make initial observations of the original condition of the following materials to be tested; record your observations in Table 5-1.

Name ____________________ **Date** ____________

Table 5-1. **Cup #** __________ **Soil Type** ____________________

Material	General Appearance and Condition	Color	Texture
Paper			
Plastic			
Biodegradable plastic			
Natural fiber			
Synthetic fiber			
Dried plant material			

Name ______________________________ **Date** ______________

Table 5-2. **Cup #** ______________ **Soil Type** ______________

Fiber	At Start	First Week	Second Week	Third Week	Fourth Week	Fifth Week	Sixth Week	Seventh Week
Natural (Cotton)	+++							
Synthetic (Polyester)	+++							

Rate fiber strength as follows:

+++ Original strength at start

++ Starting to show signs of disintegration, such as breaks or cuts of the thinner fibers that make up the sample

\+ Very weak, but still held together

– Disintegrated or decomposed

1. paper strips
2. plastic strips
3. biodegradable plastic strips
4. natural fibers
5. synthetic fibers
6. dried plant material

Note: Other materials can also be tested for biodegradation.

b. Test the original strength of the fiber (synthetic and natural) by stretching the fiber; record fiber strength using the key provided in Table 5-2.

c. Insert a piece of each of the materials to be tested into the soil of each cup, covering about half the surface area of each material. A portion of each material tested should be exposed above the soil surface. The exposed portions can be used to locate the buried materials and as reference guides to the original conditions of the materials.

d. Press the materials lightly into the soil and cover each cup with a clear plastic top.

e. Place your cups on a tray and store them at room temperature.

Observations

1. Record your observations of the original condition of the materials in Table 5-1.

   Note: Ask your instructor for multiple copies of Table 5-1 for use throughout the experiment.

2. Record your observations of the original strength of the fibers in Table 5-2.

3. Observe your setup once a week and record your results in Table 5-1. Remove the materials one by one, gently clean off the soil material, and examine each carefully with a 10X magnifying glass. Look for differences in general characteristics, such as size, shape, texture, breaks in the fibers, and fading of colors.

4. Test the strength of the fibers (synthetic and natural) once a week and compare these results to those of the original fibers. Record your observations in Table 5-2.

Conclusions

1. Which of the soil materials provided for the best biodegradation? Explain your answer.

   ____

   ____

   ____

   ____

   ____

2. When did you first notice signs of degradation?

   ____

   ____

   ____

   ____

3. What difference was there in the degradation between the natural fiber and the synthetic fiber?

4. Which of the cups had the most degrading activity? Explain why you think it did.

5. Which material decomposed the fastest? Why do you think it had the highest rate of degradation?

SECTION TWO

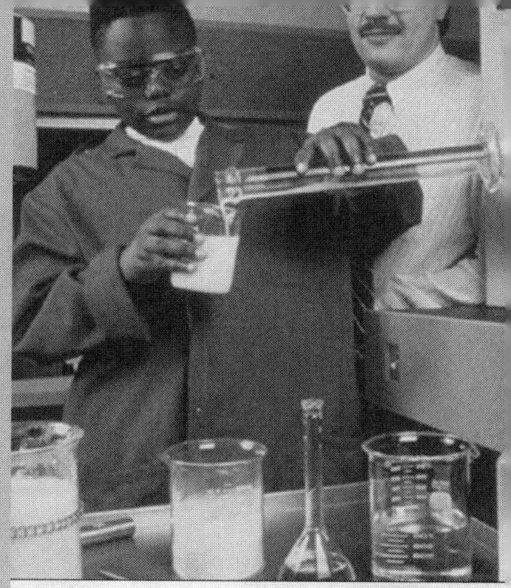

You and the New Millennium

EXERCISE 6: Get a Job—Your Career in Agriscience

Materials Needed

- ✓ Pencil
- ✓ Pen or marker
- ✓ Thumbtack
- ✓ Sheet of posterboard—18 inches × 24 inches
- ✓ Scissors
- ✓ Glue
- ✓ *Occupational Outlook Handbook*
- ✓ *Dictionary of Occupational Titles*
- ✓ Internet access

Purpose

To introduce the different agriscience careers.

Procedure

1. Preparing the career wheel

a. Using a pair of scissors, cut out the career wheel.

b. Cut out a piece of posterboard the same size as the career wheel.

c. Glue the career wheel to the cut out piece of posterboard.

d. Cut a rectangular piece of posterboard to measure 8½ inches by 11 inches.

e. Place the thumbtack in the middle of the career wheel and fasten it to the 8½-by-11-inch piece of posterboard.

f. Cut out an arrow from the posterboard. Place it on the 8½-by-11-inch piece of posterboard pointing toward the wheel.

g. When you finish, your career wheel should look like the diagram below.

Career Wheel
Agriscience Processing Products and Distribution
Horticulture
Agricultural Mechanics
Agriscience Professions
Agriscience Supplies and Services
Forestry
Renewable Natural Resources
Production Agriculture

2. Choosing a career.

a. Spin the career wheel. The arrow will point to one of the career areas.

b. Select a job for the career area from the list found under Procedure 3.

c. Using the *Occupational Outlook Handbook, Dictionary of Occupational Titles,* or the Internet, complete Observation 1 for the job you selected.

d. Spin the wheel again and repeat procedures 2b and 2c to complete Observation 2.

e. Without using the wheel, select a job of your choice from the lists under Procedure 3. Complete Observation 3 for the job you selected.

3. Agricultural career areas and jobs listed for each:

a. Dairyman

Crop Farmer
Fruit and Nut Farmer
Livestock Producer
Other List ____________
Vegetable Farmer
Horticultural Grower
Poultry Producer

b. Agriscience Processing, Products, and Distribution

Ag Establishment Inspector
Cattle Buyer
Cotton Grader
Federal Grain Inspector
Food Processing Supervisor
Fruit and Vegetable Grader
Fruit Press Operator
Grain Broker
Grain Elevator Operator
Livestock Commission Agent
Meat Inspector
Milk Plant Supervisor
Produce Commission Agent
Tobacco Buyer
Other List ________________
Butcher
Christmas Tree Grader
Farm Stand Operator
Food and Drug Inspector
Fruit Distributor
Flower Grader
Grain Buyer
Hog Buyer
Livestock Yard Supervisor
Meatcutter
Produce Buyer
Quality Control Supervisor
Weights and Measures Official
Winery Supervisor
Wool Buyer

c. Horticulture

Floral Designer
Florist
Greenhouse Manager
Horticulturist
Landscape Architect
Nursery Operator
Turf Farmer
Other List ________________
Floral Shop Operator
Golf Course Superintendent
Greenskeeper
Hydroponics Grower
Landscaper
Plant Breeder
Turf Manager

d. Forestry

Forester
Heavy Equipment Operator
Logging Operations Inspector
Nursery Operator
Plant Breeder
Tree Surgeon
Forest Ranger
Log Grader
Lumber Mill Operator
Park Ranger
Timber Manager
Other List ________________

e. Renewable Natural Resources

Animal Behaviorist
Animal Ecologist
Animal Taxonomist
Environmental Conservation Officer
Environmentalist
Fire Warden
Trapper
Forest Ranger
Game Farm Supervisor
Game Warden
Ground Water Geologist
Park Ranger
Range Conservationist
Resource Manager
Soil Conservationist
Forest Firefighter/Warden
Water Resources Manager
Wildlife Manager
Other List ________________

f. Agriscience Supplies and Service

Aerial Crop Duster
Ag Chemical Dealer
Animal Groomer
Animal Inspector
Animal Trainer
Artificial Breeding Technician
Biostatistician
Chemical Distributor
Computer Analyst
Computer Programmer
Custom Operator
Dog Groomer
Farm Appraiser
Farrier
Feed Ration Developer and Analyst
Fiber Technologist
Field Sales Representative,
Animal Health Products
Field Sales Representative, Crop Chemicals, Machinery
Kennel Operator
Meteorological Analyst
Pet Shop Operator
Poultry Hatchery Manager
Poultry Inseminator
Salesperson
Sheep Shearer
Ag Aviator
Ag Equipment Dealer
Animal Health Products Distributor
Animal Keeper
Artificial Breeding Distributor
Artificial Inseminator
Chemical Applicator
Computer Operator
Computer Salesperson
Dairy Management Specialist
Farm Auctioneer
Feed Mill Operator
Fertilizer Plant Supervisor
Field Inspector
Field Sales Representative, Agricultural Equipment
Harness Maker
Harvest Contractor
Horse Trainer
Insect and Disease Inspector
Lab Technician
Pest Control Technician
Poultry Field Service Technician
Sales Manager
Service Technician
Other List ______________

g. Agricultural Mechanics

Ag Construction Engineer
Ag Equipment Designer
Ag Safety Engineer
Ag Electrician
Ag Plumber
Diesel Mechanic

Equipment Operator
Irrigation Engineer
Machinist
Research Engineer
Soil Engineer
Other List ___________
Hydraulic Engineer
Land Surveyor
Parts Manager
Safety Inspector
Welder

h. Agriscience Professions

Ag Accountant
Ag Association Executive
Ag Corporation Executive
Ag Educator
Ag Extension Specialist
Ag Lawyer
Ag Market Analyst
Ag News Director
Agronomist
Animal Geneticist
Animal Physiologist
Agriculturist
Bacteriologist
Bioengineer
Botanist
Credit Analyst
Dendrologist
Embryologist
Entomologist
4-H Youth Assistant
Farm Broadcaster
Food Chemist
Graphic Designer
Horticulture Instructor
Ichthyologist
International Specialist
Land Bank Branch Manager
Magazine Writer
Marine Biologist
Media Buyer
Mycobiologist
Organic Chemist
Ova Transplant Specialist
Parasitologist
Photographer
Plant Ecologist
Plant Nutritionist
Plant Taxonomist
Poultry Scientist
Publicist
Reproductive Physiologist
Satellite Technician
Scientific Writer
Software Reviewer
Vertebrate Zoologist
Veterinary Assistant
Veterinarian Technician
Vidiculturist
Other List ___________
Ag Advertising Executive
Ag Consultant
Ag Economist
Ag Extension Educator
Ag Journalist
Ag Loan Officer
Ag Mechanics Teacher
Agriculture Attaché
Animal Cytologist
Animal Nutritionist
Animal Scientist
Avian Veterinarian
Biochemist
Biophysicist
Computer Specialist
Daily Nutrition Specialist
Electronic Editor
Environmental Educator
Equine Dentist
Farm Appraiser
Farm Investment Manager
Foreign Affairs Official
Herpetologist
Hydrologist
Information Director
Invertebrate Zoologist
Liminologist
Mammalogist
Marketing Analyst
Microbiologist
Nematologist
Ornithologist
Paleobiologist
Pharmaceutical Chemist
Plant Cytologist
Plant Geneticist
Plant Pathologist
Pomologist
Public Relations Manager
Publisher
Rural Sociologist
Scientific Artist
Silviculturist
Soil Scientist
Veterinarian
Veterinarian Pathologist
Virologist
Vocational Agriculture
Instructor/FFA Advisor

Observations

1. Find the information requested in Observation 1 by using the *Dictionary of Occupational Titles* and the *Occupational Outlook Handbook.*
2. Find the information requested in Observation 2 by using the *Dictionary of Occupational Titles* and the *Occupational Outlook Handbook.*
3. Find the information requested in Observation 3 by using the *Dictionary of Occupational Titles* and the *Occupational Outlook Handbook.*

Conclusions

1. How does the nature of work differ for each of the three jobs? How is it similar?

____

____

____

____

2. How do the working conditions differ for each of the three jobs? How are the conditions the same?

____

____

____

____

3. How do the qualifications for each of the three jobs differ? How are the qualifications the same?

____

____

____

____

4. How does the job outlook compare for each of the three jobs?

____

____

____

____

5. How do the earnings compare for each of the three jobs?

____

____

____

____

OBSERVATION 1

Career Wheel Challenge

Career Area ______ **Job Selected** ______

Nature of the work:

Working conditions:

Training and other qualifications:

Job outlook:

Earnings:

OBSERVATION 2

Career Wheel Challenge

Career Area ______ **Job Selected** ______

Nature of the work:

Working conditions:

Training and other qualifications:

Job outlook:

Earnings:

OBSERVATION 3

Career Wheel Challenge

Career Area ____________ **Job Selected** ____________

Nature of the work:

Working conditions:

Training and other qualifications:

Job outlook:

Earnings:

Which Career Will Be a Winner for You!?

Name ______________________________ Date ______________

Using your *Agriscience* textbook, turn to Figures 4-6, 4-8, 4-9, 4-11, 4-13, 4-15, 4-17, and 4-19 and select one career that interests you and write that career on the bracket located on the left side of this activity sheet. (Choose two careers from Figure 4-19). *Make sure each career is different.* Next, select one career from the top two left brackets and write the winner on the line connected to the top two brackets. Continue the process for the middle and bottom sets of brackets. Continue the same process until you have one career written on the single right bracket. That's YOUR winning career!!

EXERCISE 7: Experience Required—Your SAEP

Materials Needed

- ✓ *Official FFA Manual*
- ✓ SAEP Record Book
- ✓ Pen or pencil

Purpose

To help in planning a Supervised Agricultural Experience Program (SAEP).

Procedure

1. Planning your SAEP
   a. List the various agricultural crop or livestock enterprises produced successfully in your community (Observation 1). Circle those that interest you.
   b. List the various agribusinesses found in your community that support production agriculture (Observation 2). Circle those that interest you.
   c. List improvement activities you could carry out (Observation 3).
   d. List agriscience skills you could perform (Observation 4).
   e. Look through your SAEP Record Book. What type of information would you record in it (Observation 5)?
   f. Complete the Resources Inventory (Observation 6).
   g. Complete the "Selecting a Supervised Agricultural Experience Program" form (Observation 7).
   h. Optional: Visit the SAEP of an outstanding agriscience junior or senior or an American FFA Degree recipient (if one is nearby). Ask how he or she got started.
2. SAEP awards
   a. Use your *Official FFA Manual* to identify individual awards for which you could apply based on your proposed SAEP. List these in Observation 8.
   b. Ask your agriscience instructor to help you list other state and local awards for which you can apply (Observation 9).

Observations

1. Crop and livestock enterprises produced in my community: ____________________
____
____
____

2. Agribusinesses found in my community: ____________________
____
____
____

3. Improvement projects (applied activities) that I can do: ____________________
____
____
____

4. Supplementary skills (applied activities) that I can perform: ____________________
____
____
____

5. Information that I would record in my SAEP record book: ____________________
____
____
____

6. Complete the Resources Inventory.

7. Complete the “Selecting a Supervised Agricultural Experience Program.”

RESOURCES INVENTORY

1. Name ____________________ Age __________ Class __________
2. Address ____________________ Phone ____________________
3. Parents' names ____
4. Occupations ____
5. Number of brothers and sisters ____________________
6. I live on a farm __________ in town __________ on an acreage __________
7. Is land available for you to rent to produce crops? Yes __________ No __________
   a. If yes, how many acres? __________ b. Which crops? __________
   c. Location of the land ____
8. Are facilities available for you to rent to produce livestock or livestock products?

   Yes __________ No __________
   a. If yes, what type? ____
   b. Number __________ c. Location of facilities ____________________
9. Do you have space available for a garden? Yes __________ No __________
10. Do you have a greenhouse available for your use? Yes __________ No __________
11. Do you have facilities for doing mechanical work? Yes __________ No __________
12. Would you be interested in producing livestock on the school farm?

    Yes __________ No __________ If yes, what type of livestock? __________
13. Would you be interested in producing crops on the school farm?

    Yes __________ No __________ If yes, what crop? ____________________
14. Do you have an agricultural job available to you?

    Yes __________ No __________ Describe ____________________

    ____

SELECTING A SUPERVISED AGRICULTURAL EXPERIENCE PROGRAM

for

____

Name of Student

Instructions: Use this form to tentatively decide on a beginning agriscience SAEP. This information will be used in agriscience classes to develop detailed plans for obtaining experiences.

My interest in agriscience is ____

____

____

Based on my interest and the opportunities available to me to get practical experience in agriscience, I plan to include the following in my agriscience SAEP:

1. Production or productive enterprises (examples: beef, dairy, nursery production, Christmas trees)

   ____

   ____

2. Placement in an agribusiness (examples: supply store, florist shop, nursery, golf course, landscape contracting)

   ____

   ____

3. Improvement (examples: landscape your home, fertilize your lawn, plant trees)

   ____

   ____

4. Agriscience skills (examples: change spark plugs, weld, change the oil in small engines)

   ____

   ____

5. Other activities (examples: projects on school facility)

   ____

   ____

Conclusions

1. National individual awards for which I can apply. ________________________
Obtain the paperwork from your instructor, and apply for the awards you qualify for.

2. State or local awards for which I can apply. ____________________________
Obtain the paperwork from your instructor, and apply for the awards you qualify for.

EXERCISE 8: Show Me the Way— Leadership in Agriscience

Materials Needed

- ✓ *Official FFA Manual*
- ✓ *Parliamentary Guide for FFA* or *Robert's Rules of Order*
- ✓ Pen or pencil

Purpose

To introduce the procedures for conducting FFA meetings and the rules of parliamentary law.

Procedure

1. Conducting FFA meetings
   a. Study the following diagram of the FFA meeting room. Learn where each of the officers is stationed and the symbol of each office.

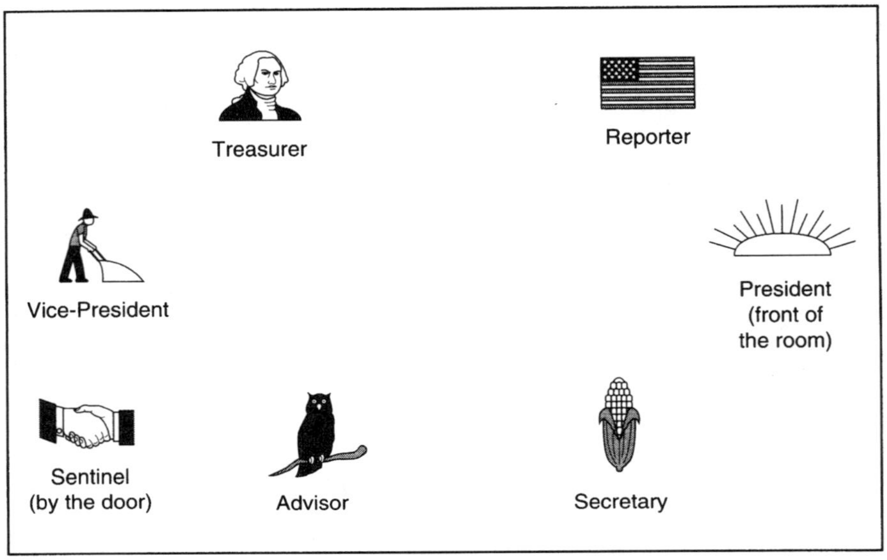

   b. In the *Official FFA Manual,* read over the opening and closing ceremonies for all FFA meetings.
   c. Let each class member read a part in the ceremonies. Take turns. Go over the ceremonies several times.
   d. Select one officer's position, and write the part in the ceremonies (Observation 2).

2. Learning parliamentary law
   a. In the *Official FFA Manual,* study the order of business for a meeting. Complete Observation 3.
   b. Study the classifications, names, purposes, and pertinent information for the primary motions used in conducting FFA business. Use the *Parliamentary Guide for FFA* or *Robert's Rules of Order.* Complete Observation 4.
   c. Practice conducting a meeting using the sample business found in Observation 5. Fill in the blanks appropriately.

Observations

1. Complete the following meeting room diagram by listing the officers and the symbols where they should be stationed.

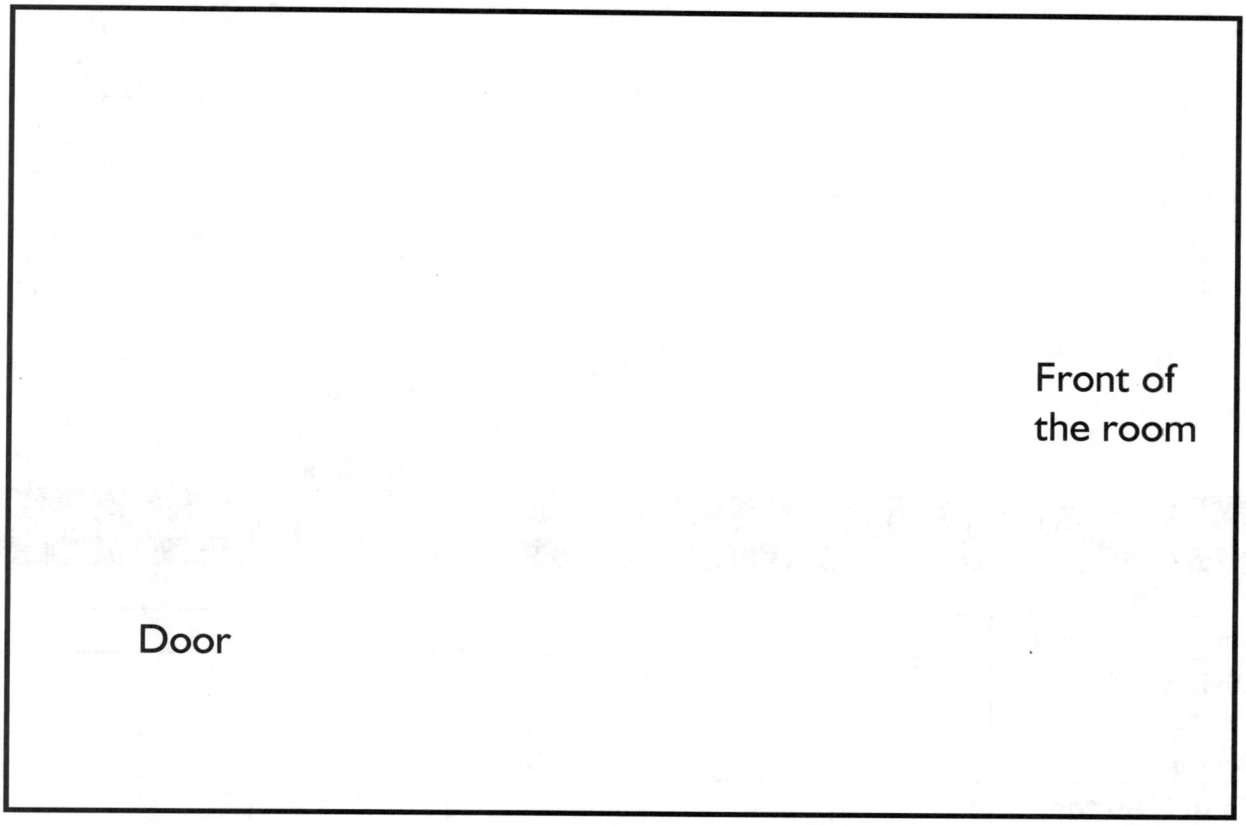

2. Write one part in the opening ceremony. ____

____

____

____

____

____

____

____

____

____

____

____

____

____

3. List the order of business for an FFA meeting. ______________________________

4. Complete the blanks in the following chart.

		Yes or No	Yes or No	Majority Two-thirds or None	Yes or No
Name of Motion	**Kind of Motion**	**Debatable**	**Amendable**	**Vote Required**	**Second Required**
Main Motion					
Lay on the Table					
Division of the House					
Amend					
Take from the Table					
Refer to a Committee					
Previous Question					
Reconsider					
Parliamentary Inquiry					
Rescind					
Suspend the Rules					
Adjourn					

5. List the information requested for each motion. ______________________________

Practicing Parliamentary Law

1. A member moves to paint the interior of the agriculture building. Write an introductory statement; include proper terminology. ______

2. The motion is debated by two or more members. Write down the discussion for and against the motion. ______

3. Lay the amended motion on the table. Write an introductory statement; include proper terminology. ______

4. A motion is made to have the annual parent–member banquet during National FFA Week in February. Write an introductory statement; include proper terminology.

5. A member moves to refer the motion to a committee. Write an introductory statement; include proper terminology. ______________________________

6. A member discusses the motion to refer to a committee. Write discussion for the motion to refer. ______________________________

7. A member moves the previous question. Write the proper terminology to make this motion. ______________________________

8. A member moves to adjourn the meeting. Write the proper terminology for this motion.

Conclusions

1. How can learning leadership in agriscience help you to become a better citizen in your community?

2. In what other meetings or organizations can you use parliamentary law?

3. List five individuals in your community whom you consider to be leaders. Give reasons for selecting each.

SECTION THREE

Natural Resources Management

EXERCISE 9: Watt's the Big Deal—Creating Electricity Out of Thin Air (An exercise adapted from a National Renewable Energy Laboratory Booklet)

Materials Needed

- ✓ 100 cm of copper wire
- ✓ 1 bar magnet
- ✓ 1 electrical meter
- ✓ 1 cardboard tube
- ✓ Duct tape
- ✓ Fan or turn table

Purpose

The demand for electricity is at an all time high, and the price for it is also at its peak. Some farmers have begun to add wind energy to the list of commodities they produce on their land. To create wind energy, large turbines are strategically placed in known windy areas. The wind catches the turbine's blades and creates lift causing them to rotate. The blades are connected to a drive shaft that turns an electrical generator. In this lab, you will explore how electricity is created in an electrical generator.

Procedure

1. Wrap the copper wire around the cardboard tube about 20 to 25 times.
2. Connect both ends of the wire to the electrical meter (one end to each probe). The meter will register any electricity created. Movement on the meter will indicate electrical activity.
3. Hold the magnet and move it around the tube.
4. Next take the magnet and move it through the tube. Try doing this at different speeds.
5. Place the magnet on a flat surface and move the tube over it.
6. Attach either the tube or the magnet onto a turn table or fan to find the position that creates the most electricity (optional).

Observations

Procedure	Meter Reading Range
Procedure 3	
Procedure 4	
Procedure 5	
Procedure 6 (opt.)	

Conclusions

1. Describe how you were moving both the tube and the magnet when the electrical meter was showing the most electrical activity (needle movement or highest readings).

2. When did the meter register more activity[[[…]]]when the magnet was being moved around it or through it?

3. If a turn table or fan was used, did the speed it provided, create more energy?

4. Using the information from question two, draw a sketch of a wind turbine that would produce the most electricity.

EXERCISE 10: Getting Down and Dirty—Preventing Soil Erosion

Materials Needed

- ✓ Six plant trays (or shoe boxes cut to 5 inches deep)
- ✓ Plastic sheeting to line trays
- ✓ Six sprinkling cans
- ✓ Six beakers
- ✓ Trowel
- ✓ Scissors
- ✓ Stopwatches
- ✓ Soil
- ✓ Sod (grass-covered soil from a lawn)
- ✓ Water
- ✓ Supports to form inclines for trays

Purpose

To compare the rates and amounts of erosion that result from different land uses.

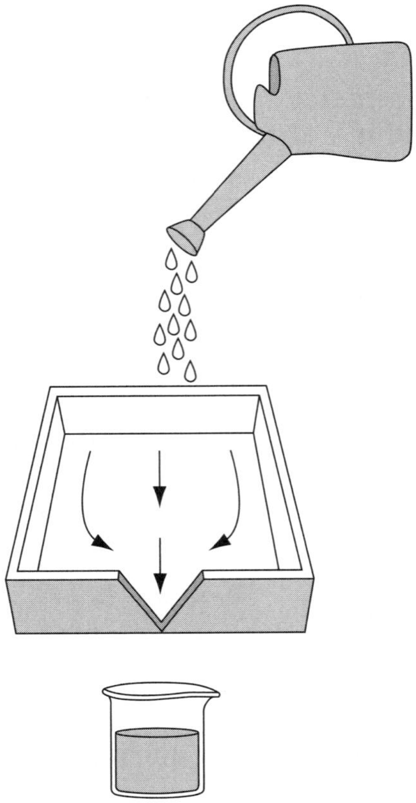

Procedure

1. Preparing the materials

a. Cut a "V" notch in one end of each plant tray (or shoe box).

b. Line each tray with plastic, leaving an overhang of about 2 inches at the notch. This will serve as a spout.

c. Number and label each tray as you fill each of the six trays differently as follows:

1. moist soil packed firmly (bare soil)
2. sod (cover crop)
3. moist soil packed firmly; make packed furrows running across the tray's width (contouring)

4. moist soil packed firmly; make packed furrows running down the length of the tray (plowing up and down a hill)
5. moist soil packed firmly; form steps across the width of the tray with the trowel (terracing)
6. alternate three strips of moist packed soil with three strips of sod (strip cropping)

d. Position the trays on supports so the trays are tilted at an equal incline. Place beakers under the plastic overhangs at the "V" notches (spouts).

2. Conducting the experiment

a. Measure an equal amount of water into each sprinkling can.

b. With the help of fellow students, hold the sprinkling cans about 1 foot over the high point of each tray. Pour water steadily for 5 seconds onto each tray simultaneously.

c. In Observation 1, record the amount of time that water flows from the spout of each box.

d. After the water finishes draining, record the amount of runoff in each beaker in Observation 2.

e. Let the runoff settle, and then measure the volume of sediment in each beaker. Record the results in Observation 3.

f. Arrange the trays to increase the slope and repeat procedures 2a through 2e, recording the results in Observations 4 through 6.

Observations

1.

	Tray 1	Tray 2	Tray 3	Tray 4	Tray 5	Tray 6
Amount of Time Water Flows from the Spout						

2.

	Beaker 1	Beaker 2	Beaker 3	Beaker 4	Beaker 5	Beaker 6
Amount of Runoff						

3.

	Beaker 1	Beaker 2	Beaker 3	Beaker 4	Beaker 5	Beaker 6
Volume of Sediment						

4.

	Tray 1	Tray 2	Tray 3	Tray 4	Tray 5	Tray 6
Amount of Time Water Flows from the Spout at Increased Slope						

5.

	Beaker 1	Beaker 2	Beaker 3	Beaker 4	Beaker 5	Beaker 6
Amount of Runoff at Increased Slope						

6.

	Beaker 1	Beaker 2	Beaker 3	Beaker 4	Beaker 5	Beaker 6
Volume of Sediment at Increased Slope						

Conclusions

1. Which tray had the most runoff? Which tray lost the most soil?

2. Which tray's content best prevented erosion?

3. Is there a relationship between the time it took for each tray to finish draining, and how much soil washed off? Explain your answer.

4. How does the slope affect the soil loss?

5. How would you apply what you have learned from this experiment to soil conservation?

EXERCISE 11: Talking Dirty—Soil Fertility

Materials Needed

- ✓ Seven identical 6-inch flower pots with drainage outlets at the bottom
- ✓ Pot chip or screen
- ✓ Vermiculite or perlite
- ✓ Mixing bin or box
- ✓ Pea and tomato seeds
- ✓ Nitrogen source—NH_4NO_3
- ✓ Phosphorus source—superphosphate
- ✓ Potassium source—KCl
- ✓ Calcium source—$CaCO_3$
- ✓ Materials for labeling pots
- ✓ Laboratory balance or scale
- ✓ Ruler or yardstick
- ✓ Personal Safety Set (includes rubber gloves and safety goggles or glasses)

Purpose

To understand the value of soil nutrients by observing the effects of a deficiency of certain nutrients in plants.

Procedure

1. Place a screen or pot chip over the hole in the bottom of each pot.
2. Fill each pot with soil to 1 inch from the top.
3. Label the pots to correspond with the names of the treatments in column one of the following chart. One pot should be labeled for each treatment.

Treatment	*Nitrogen*	*Phosphorus*	*Potassium*	*Calcium*
No Fertilizer	None	None	None	None
Complete Fertilizer	1.1 g NH_4NO_3	1.25 g Superphosphate	.3 g KCl	None
Complete Fertilizer with Calcium	1.1 g NH_4NO_3	1.25 g Superphosphate	.3 g KCl	1.5 g $CaCO_3$
Nitrogen Deficient	None	1.25 g Superphosphate	.3 g KCl	None
Phosphorus Deficient	1.1 g NH_4NO_3	None	.3 g KCl	None
Potassium Deficient	1.1 g NH_4NO_3	1.25 g Superphosphate	None	None
Calcium Only	None	None	None	1.5 g $CaCO_3$

4. Weigh the amounts of fertilizer specified in columns 2, 3, 4, and 5 of the preceding chart. Dump the soil into a bin or box and mix thoroughly with the fertilizer before returning the soil to the pot. Be sure not to mix materials from separate pots. Wash your hands after handling fertilizer.
5. Divide each pot into half with an imaginary line. Plant five pea seeds in one half and five tomato seeds in the other half. Water them in.

6. Place the pots in a location with adequate light and temperature for growing. Ensure that all pots have the same growing conditions.

7. Check the plants at regular intervals and water them as needed.

8. When the seeds have germinated and the plants are established, thin the plants so that each pot has the same number of each kind of plant.

9. Observe the plants each week, making note of differences in color, height, width, and any other differences between pots. Record this information in the table in Observation 1.

10. At the conclusion of the experiment, harvest all plants by pulling them up by the roots. Weigh all the plants of the same kind in each pot together and record the weights on the chart in Observation 2.

11. Let the plants dry in the sun for 4 days. Be sure to keep the plants from different pots separated. Weigh each kind of plant in each pot again separately. Record the dry weights on the chart in Observation 2.

Observations

1. In the following table, record your observations of rate of growth, color, and so on.

Treatment							
Observations	No Fertilizer	Complete Fertilizer	Complete Fertilizer with Calcium	Nitrogen Deficient	Phosphorus Deficient	Potassium Deficient	Calcium Only
Week 1							
Week 2							
Week 3							
Week 4							
Week 5							
Week 6							
Week 7							
Week 8							

2. In the following chart, record the harvest weight for each kind of plant for each treatment.

Treatment	Green Weight		Dry Weight	
	Peas	Tomatoes	Peas	Tomatoes
No Fertilizer				
Complete Fertilizer				
Complete Fertilizer with Calcium				
Phosphorus Deficient				
Potassium Deficient				
Calcium Only				

Conclusions

1. What were the signs of deficiencies in the plants you grew, if any?
   a. Nitrogen ______________________________
   ____
   b. Phosphorus ______________________________
   ____
   c. Potassium ______________________________
   ____
   d. Calcium ______________________________
   ____

2. Which plants seemed to grow best in the soil to which calcium was added? __________
____

3. How did the green weights compare to the dry weights? ______________________________
____
____
____

EXERCISE 12: Hold the Soil Please—Growing Plants without Soil

Materials Needed

- ✓ Hydroponics Experiment Kit
- ✓ Personal Safety Set

SAFETY NOTE

Wear eye protection glasses when handling any chemical solution. Handle chemical reagents with care, and follow your instructor's directions. Avoid contact with skin or eyes.

Purpose

To demonstrate the art of growing plants without soil (hydroponics).

Procedure

1. Fill the transparent plastic culture chamber with deionized water to the ridge located near the top edge.
2. Add 50 drops of each of the following nutrient solutions to the water-filled chamber and stir with a clean glass rod.
   a. $Ca(NO_3)_2$ solution (calcium nitrate)
   b. KNO_3 solution (potassium nitrate)
   c. $MgSO_4$ and KH_2PO_4 solution (magnesium sulphate and potassium phosphate)
3. Add 5 drops of the iron chelate solution to the chamber.
4. Add 5 drops of the trace elements solution to the chamber.
5. Carefully place a seed inside the slit of a germinating plug. This should then be inserted into a hole provided in the white germinating float. When each of the 12 holes has been filled with seeds (and plugs), carefully place the float into the chamber so that all plugs are in contact with the liquid in the chamber.
6. Replace the opaquing shield on the lower portion of the culture chamber, if it has been removed. The shield will inhibit the growth of algae in the solution and should be removed to observe root growth.
7. Replace the vented cover and allow the plants to germinate. A south or west window is the best location for the culture chamber.
8. As the solution is used by the plants, it should be brought back to the original level by adding deionized water. After the seeds have germinated, replace the nutrients and trace elements once a week.
9. Observe the growth of the plants daily for 2 weeks and record the progress on the chart in Observation 1.

Observations

1. Record the daily progress of the plants in the following chart.

Day	Plant Growth Conditions			
	Length of Shoots and Roots	Coloration of Roots	Coloration of Leaves	Other Conditions
1				
2				
3				
4				
5				
6				
7				
8				
9				
10				
11				
12				
13				
14				

Conclusions

1. How can the plants germinate without soil?

2. Did you observe any deficiencies in the plants?

3. What color were the roots of the plants?

4. What possibilities do you see for the use of hydroponics in the future?

EXERCISE 13: There's Something Fishy Here—Fish Anatomy

Materials Needed

- ✓ Dissection Lab Perch Kit
- ✓ Dissection set
- ✓ Magnifying glass
- ✓ Personal Safety Set

SAFETY NOTE

Always wear eye protection glasses and rubber gloves when handling dissection specimens. Wash hands before leaving the laboratory.

Purpose

To become familiar with the major internal and external features and structures of the perch and with their functions.

Procedure

1. Study and become familiar with the following terms:
   a. **dorsal**—back or upper
   b. **ventral**—abdominal or lower
   c. **anterior**—forward or front
   d. **posterior**—hind or rear
   e. **median**—toward the middle
2. Refer to Chapter 12 of *Agriscience: Fundamentals & Applications, 5th Edition,* to study the parts and learn their functions as defined.

   External Anatomy

   a. **Mouth**—at anterior end of the animal. The position of the mouth reflects the manner whereby the fish feeds. The fish possessing a terminal mouth feeds by overtaking prey while swimming. The fish possessing a dorsal mouth is usually a surface feeder. If the mouth is ventral, the fish is usually a bottom feeder.
   b. **Nostril**—two pairs, each pair opens into an olfactory (smell) chamber and allows water to pass through the chamber. The movement of water is essential to odor perception.
   c. **Eye**—two large spherical eyes are located on each side of the head.
   d. **Gill openings**—allow water passing through the gills to exit.
   e. **Operculum**—a semicircular flap that covers the gills.
   f. **Pectoral fins**—paired fins found on either side of the perch.
   g. **Pelvic fins**—paired fins found close together on the ventral surface of the fish trunk.

h. **Anterior dorsal fin**—single fin found on the dorsal side of the fish and supported by 13 to 15 spiny rays.

i. **Posterior dorsal fin**—single fin located directly behind the anterior dorsal fin and supported by 7 to 8 soft rays.

j. **Caudal fin**—the fleshy end of the tail is surrounded by this fin; the type of caudal fin found on the perch is called *homocercal.*

k. **Anal fin**—single fin located behind the anus on the ventral side and supported by soft rays.

l. **Anus**—opening through which feces are eliminated from the fish body.

m. **Skin**—envelope for the body and the first line of defense against disease; it also affords protection from, and adjustments to, environmental factors that influence life.

n. **Mucous glands**—glands located in the skin, the secretions of which give the slimy touch and odor to the fish.

o. **Fish scales**—thin, flexible plates that provide a flexible "armor" to protect the fish. They are used in growth and age studies. The scales of perch are called *ctenoid* because the exposed portion bears tiny, spiny structures called *cteni* (teeth).

p. **Lateral line system**—a sensory organ system restricted to the fishes and aquatic stages of amphibians; it enables a fish to detect localized disturbances from currents or vibrations caused by fixed or moving objects.

3. Study external anatomy

a. Place the perch on the dissection tray, side down, with the head facing to the left, dorsal fins up.

b. Locate and identify the external features of the perch. Label these on the diagram in Observation 1.

c. Using forceps, carefully remove a fish scale from the trunk region and examine it under a magnifying glass. Identify the cteni. Complete Observation 2.

4. Study the internal parts of the perch found in text Chapter 12. Each part's function is discussed in the appropriate section under "Internal anatomy."

Internal Anatomy

a. **Muscular system**—controls body movements.

1. With a scalpel, carefully cut a large (2-inch) square window on the lateral surface of the perch directly below the anterior dorsal fin.
2. Using forceps, carefully peel away the skin to expose the muscle layer beneath. Identify and label the trunk muscles on the diagram in Observation 3. Trunk muscles are attached to the skeleton and are of primary use in producing body movement.

b. **Respiratory system**—serves in the exchange of gases with the environment.

1. Using scissors, carefully cut the operculum at its base on the head to expose the gills. Observe the figure in the kit.
2. Extend the cut posteriorly from the base of the angle of the jaw around the gills. Raise and remove the operculum.
3. Pry open the mouth with a probe and observe the mouth and pharynx (pharyngeal cavity). Feeding is accomplished by sucking prey into the mouth, unlike those fishes that bite their prey.
4. Identify the following parts of the respiratory system on the fish and label them on the diagram in Observation 3.
   a. **Palate**—solid roof extending from the mouth cavity to the pharynx.
   b. **Gills**—organs of respiration that have thin walls that are supplied with blood capillaries, as are our lungs.
   c. **Gill arches**—serve as attachments for the gills.
   d. **Gill filaments**—tiny fingers running at right angles to the length of each gill. Oxygen dissolved in water diffuses through the thin wall of each gill filament into capillaries, and carbon dioxide diffuses in the opposite direction.
   e. **Gill rakers**—bony structures located on each gill arch and pointing toward the pharynx. Collectively, they act as a strainer preventing food and other particles from being carried across the gills, which could cause injury to the gills.

c. **Circulatory system**—the transport system of the body supplies oxygen and nourishment to body cells and removes metabolic wastes.

1. Remove the exposed pectoral, pelvis, and anal fins.

2. Using scissors, cut through the lateral side of the body wall.

3. Remove the section of body wall cut.

4. Hold the perch in place by pinning the head, tail, and fins to the dissection tray.

5. Identify the following parts of the circulatory system on the fish and label them on the diagram in Observation 3.
   a. **Heart**—a two-chambered, muscular organ that receives blood from the body, circulates it to the gills for oxygenation to the body, and then back to the heart. Blood in the fish passes from the atrium to the ventricle without being oxygenated.
   b. **Arteries**—carry blood from the heart to the body tissues.
   c. **Veins**—carry blood to the heart from the body tissues.
   d. **Spleen**—a related circulatory organ; an elongated structure lying on the posterior dorsal surface of the stomach. It is concerned with the production and maintenance of blood cells.

d. **Digestive system**—functions to both mechanically and chemically break down food into simple components that can be easily absorbed by the blood and passed throughout the body to provide body cells with nourishment.

1. Identify the following parts of the digestive system on the fish and label them on the diagram in Observation 3.
   a. **Esophagus**—a short, straight tube leading from the mouth to the stomach.
   b. **Stomach**—receives food from the esophagus and digests the food's protein components.
   c. **Intestinal diverticula**—three blind sacs located at the anterior-most part of the duodenum,

where it connects to the stomach. The pancreatic and bile ducts open here.

d. **Duodenum**—partially digested food is passed from the stomach to this organ for final digestion. The duodenum forms an S-shaped loop from its origin at the stomach.

e. **Intestine**—final digestion and absorption of food from the duodenum takes place here.

f. **Fat bodies**—white masses located around the intestine; they function to store food reserves.

g. **Anus**—outside opening of the intestine through which solid waste is passed.

h. **Digestive glands**

1. **Liver**—reddish-brown, lobed organ situated just anterior and dorsal to the stomach. It produces bile, which aids in fat digestion and the regulation of nutrient levels in the blood.
2. **Gallbladder**—greenish sac between the lobes of the liver; stores bile before release to the duodenum.
3. **Pancreas**—a small, folded organ located at the ventral border of the duodenum; manufactures pancreatic juices that further digest proteins.

e. **Excretory system**—removes metabolic waste products from the blood.

1. Identify the following parts of the excretory system on the fish and label them on the diagram in Observation 3.

a. **Kidneys**—paired, elongated masses pressed against the dorsal body wall.

b. **Collecting ducts**—tubes that drain each kidney. In the male, they connect to the urinary bladder. In the female, they connect the kidney to a urogenital sinus where the urinary bladder is incorporated into the oviduct.

c. **Urinary papilla**—external projection of the male urinary bladder. Waste products (urine) are eliminated from the body through a urinary pore. In the female, urine is eliminated through the abdominal pore.

d. **Urinary bladder**—a small sac of the most posterior, ventral area of the body cavity; functions to store urine.

f. **Reproductive system**—responsible for the production of sex cells and their delivery to produce offspring; sex organs are separate in the perch.

1. Identify the following parts of the reproductive system on the fish, and label them on the diagram in Observation 3.
   a. **Testes** (male)—paired, lobulated structures, posterior to the duodenum and stomach, and dorsal to the intestine. They produce sperm, which is transported to the outside by a duct called the *vas deferens* and eliminated through the genital pore.
   b. **Ovary** (female)—single, large sac located in the same position in the body as are the testes in the male. The ovary produces white spherical eggs, which fill the cavity to capacity in the spring.
   c. **Oviduct**—a funnel-like structure where the urinary bladder is incorporated. Eggs are expelled through the abdominal pore.
   d. **Other parts**
      1. **Gas bladder** or **swim bladder**—helps regulate depth independently of body motion; located ventral to the kidney and dorsal to the sex organs.

Observations

1. Label the external parts of the perch on the following diagram.

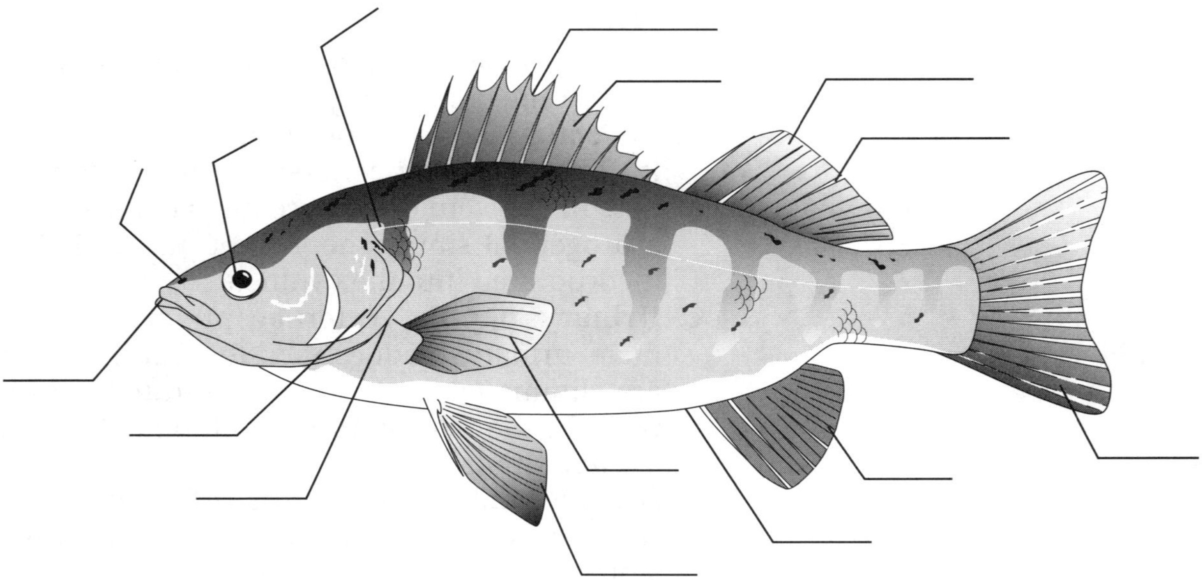

2. Draw a diagram of the fish scale and identify the cteni.

3. Label the internal parts of the perch on the following diagram.

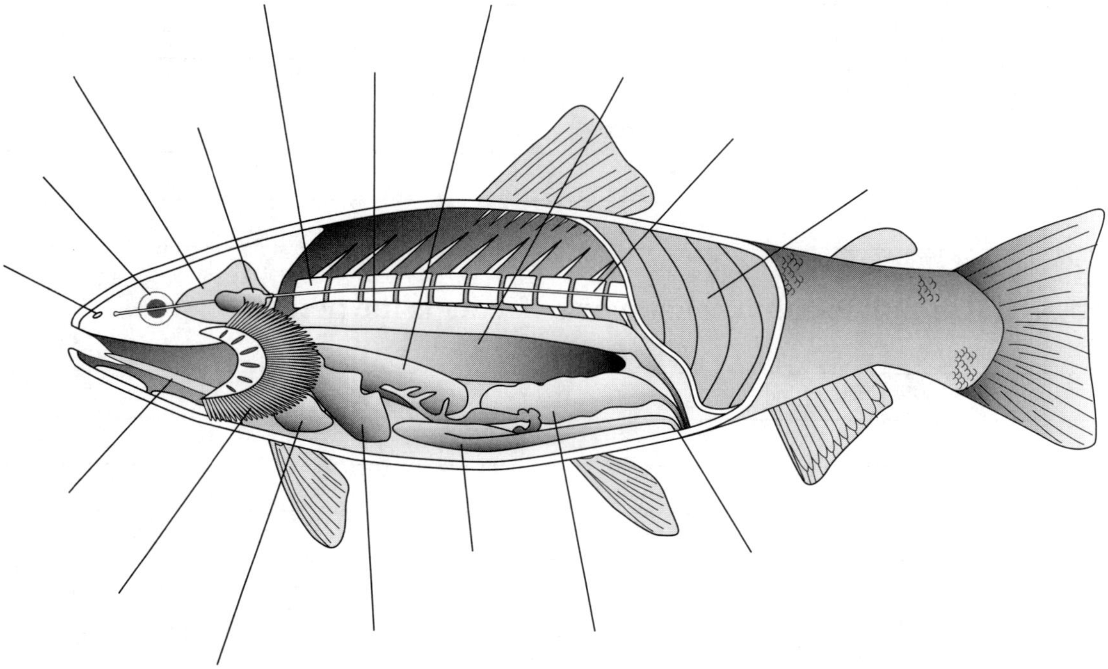

4. Is your fish male or female? How can you tell?

5. Cut open the stomach of the fish. What types of food do you find there?

Conclusions

1. How many fins does your fish have? What purpose is served by the fins?

2. How do the nostrils aid the fish in smelling?

3. Based on the mouth location, what type of feeder is the perch?

4. How is the heart of a fish different from yours?

____

____

____

____

____

5. What is the function of the gas bladder?

____

____

____

____

____

____

SECTION FOUR

Integrated Pest Management

EXERCISE 14: Are You a Competitor? Weed Identification lab

Materials Needed

- ✓ A large amount of News Paper
- ✓ Two pieces of tempered hard board, peg board, or plywood (~12" × 18" rectangles).
- ✓ Cardboard cut into ~12" × 18" rectangles
- ✓ Tie downs or bungee cords for strapping the weed portfolio tightly.
- ✓ Large pieces of construction paper for mounting the specimens.
- ✓ Plastic bags for collecting (old grocery bags work well and keep the specimens moist)
- ✓ Shovel
- ✓ Adhesive for mounting the specimens
- ✓ Blank adhesive labels
- ✓ 20 weeds collected near your home
- ✓ Internet access
- ✓ Weed Identification manuals
- ✓ Small zip style bags

Purpose

Believe it or not, but you are in competition with pests every day. Anyone who has ever weeded a garden understands that they are seriously out-numbered. Fortunately for us, integrated pest management strategies can help. The first step in dealing with pests is to know your enemy. This is the most difficult and most important part of the process. In this lab, you will be gathering common weeds found in your community and then identifying them by their common and scientific names.

Procedure

Allow several days for this lab.

1. Collect 20 weed specimens from near your home. This should be done as homework. Use a shovel to get all of the plant (roots, stem, leaves, flowers, fruit and seeds). Not all plants will have the last three items. Place the weed into a plastic sack. Leave the soil on the root ball until you are ready to press it, and then remove it.
2. While the specimen is still fresh identify it. It is much easier to identify a fresh plant than it is to identify a pressed and dried one. This can be done using the Internet (key words), weed identification guide, or a weed identification manual.
3. To mount the weed specimen, place the plant between several pages of newspaper.
   - Large plants can be bent a couple of times so it will fit on the page.
   - Show the underside of some of the leaves so that both sides of the leaves can be seen.
   - Remove as much soil as possible from the roots. Spread the roots out as much as possible.
   - Seed samples collected should be placed in a zip style bag and saved for mounting later.
4. Now that the plant is inside three or four layers of news paper, it will be placed between two pieces of cardboard. All of your samples should be layered this way and placed in-between the outer cover (constructed of lathe or thin plywood). Next, use the tie downs or bungee cords to wrap

the specimens inside the two pieces of wood. Finally place a heavy object on top of your sample portfolio for a few days.

- Change the newspaper and cardboard as often as needed (at least once a day at first). Extremely green plants are hard to preserve.

5. Once the weeds are dry, carefully attach them to a large sheet of construction paper. Take care when doing this because, the dried plants are very brittle. Place a plant identification label on the back of the page.

    - The label should include; common names, Genus and species, date, and location of collection.
    - Be sure to include the weed's seeds and fruit on the mounting page.
    - Your final portfolio should include 20 mounted plants with labels. The pages should be numbered, and they should correspond to a table of contents.

Note: In this lab, your observations and conclusions are included in the overall process.

EXERCISE 15: Don't Bug Me—Insect Anatomy

Materials Needed

- ✓ Dissection Manual Grasshopper Kit
- ✓ One pair of fine dissection scissors
- ✓ One sharp probe
- ✓ Dissecting microscope
- ✓ Personal Safety Set
- ✓ Internet access
- ✓ Text book

SAFETY NOTE

Always wear eye protection glasses and rubber gloves when handling dissection specimens. Wash hands before leaving the laboratory.

Purpose

To recognize and identify the major external and internal features and structures of the grasshopper and explain their functions.

Procedure

1. Study the information regarding anatomical orientation terminology and the biology of the grasshopper as outlined in the grasshopper kit.
2. Study the information regarding body shape and the exoskeleton as outlined in the grasshopper kit.
3. External anatomy
   a. Place the grasshopper on the dissection tray so that its head is to your left and its legs are down.
   b. Locate and identify the external features of the grasshopper by referring to the diagrams provided in the grasshopper kit.
   c. Use the dissection microscope to observe some of the smaller details.
   d. Using the text or an insect anatomy Web site, locate and label the external parts on the diagram in Observation 1.
4. Identify the sex of your grasshopper using the external anatomical features detailed in the kit. Record in Observation 2.
5. Internal anatomy
   a. Snip off the legs, wings, and antennae of the grasshopper at their bases.
   b. Using pointed iris scissors, insert one point of the scissors on the left side in the intersegmental membrane between the 10th and 11th abdominal segments just below the cercus. Cut anteriorly up the left side of the grasshopper at a level of approximately 1/8 inch above the level of the spiracles.
   c. Continue to cut at the same approximate level all the way to the head. (Observe the diagram in the kit.)
   d. Make a similar cut at the same approximate level on the right side of the insect, extending the cut anteriorly

from the 10th and 11th abdominal segment to the head. Join both cuts so that the grasshopper is completely bisected laterally.

e. Carefully lift the posterior part of the dorsal body dissection, working it toward the head, free from the larger ventral body dissection. No difficulties will be encountered until you reach the thorax. Here, cut through the dorsoventral and oblique body muscles and ligaments. In the head, free the head muscles as close to the dorsal dissection body wall as possible.

f. Remove the dorsal body dissection and place both the dorsal and ventral body dissections in the dissection tray, using insect pins to securely position both dissections.

g. Orient both dissections (ventral and dorsal) in the dissection tray with the head region facing away from you.

h. Locate and identify the internal structures of the grasshopper by referring to the diagrams provided in the grasshopper kit. Study the various systems as discussed in the kit.

i. Using an insect anatomy Web site, label the various internal parts on the diagrams in Observation 3.

Observations

1. Label the external parts of the grasshopper on the following diagram.

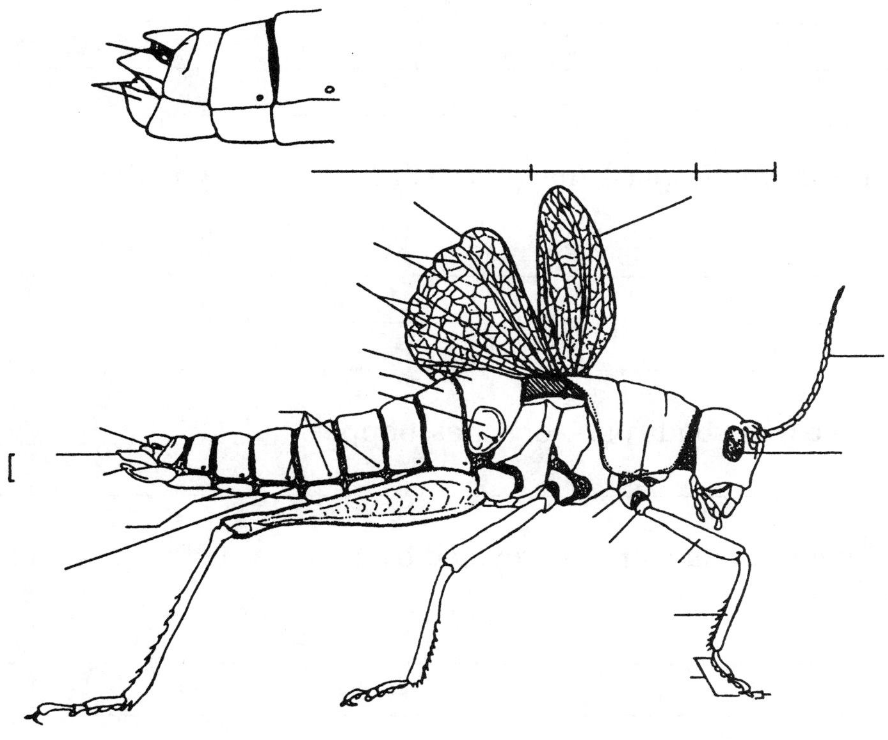

2. Circle the sex of your grasshopper. Male/Female

3. Label the internal parts of the grasshopper on the following diagrams:

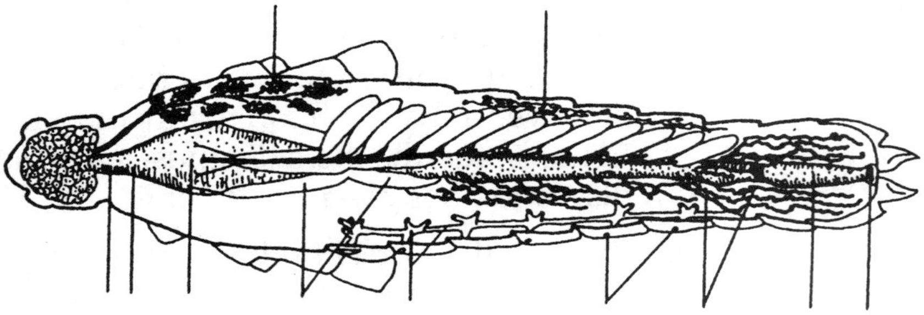

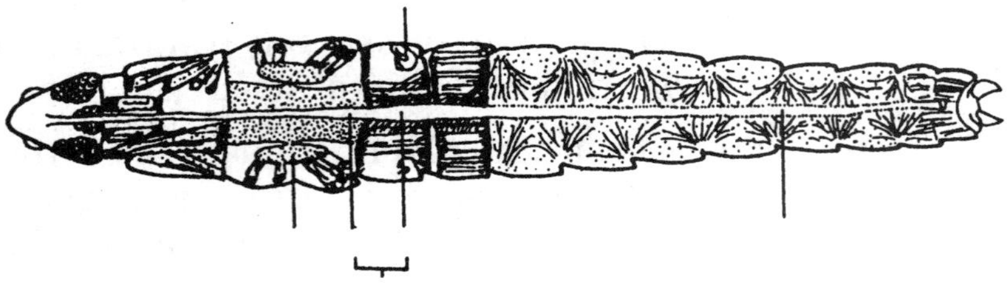

Conclusions

1. How is the exoskeleton of a grasshopper different from the skeleton of a human? How are they alike?

____

____

____

2. What three types of mouthparts may insects have? What type does the grasshopper have?

____

____

____

3. What are the three main body parts of a grasshopper? ______________________

____

4. Using the Internet or other sources provided by your instructor, explain the functions of the following:

   a. antenna ______________________________________

   b. mandibles ____________________________________

c. pretarsus ______________________________

d. spiracle ______________________________

e. ovipositor ______________________________

f. sclerites ______________________________

g. crop ______________________________

h. malpighian tubules ______________________________

i. ocellus ______________________________

j. cercus ______________________________

EXERCISE 16: Stuck on You—External Parasites

Materials Needed

- ✓ Compound microscope (used for most lice and fleas)
- ✓ Dissecting scope (sufficient for larger lice and engorged ticks)
- ✓ Insecticide powder
- ✓ Animal comb or any inexpensive plastic hair comb
- ✓ Smooth dark material (paper or plastic)
- ✓ 70 percent ethyl alcohol or isopropyl (rubbing) alcohol
- ✓ Glass jar
- ✓ Water or oil (baby oil)
- ✓ Microscope slides and cover slips
- ✓ Dissecting needle
- ✓ Personal Safety Set
- ✓ Internet access (optional)

Lice, Fleas, and Ticks: Importance to the Agriculture Industry

Purpose

External Parasites can cause discomfort and disease in domestic animals. Learning to remove these parasites will allow you to better care for your animals.

Note: When dealing with parasites and blood exposure, always wear medical gloves and eye protection glasses, and wash any area that has been exposed with an antibacterial soap.

1. In general, lice, fleas, and ticks cause

a. ______________________________

b. ______________________________

c. ______________________________

2. What diseases can fleas transmit to other animals and humans?

a. ______________________________

b. ______________________________

3. For what diseases are ticks responsible for transmitting?

____

Procedure

Note: Due to exposure issues, your instructor or a licensed animal health care professional will perform the parasite removal procedure (steps 1–4).

1. Dust the animal with an appropriate insecticide powder that will kill or immobilize the parasites. The insecticide used must be approved and safe for the breed, sex, age, and function of the animal. Carefully follow the directions on the label to minimize any harmful effects to the animal.

2. Comb the parasites out of the coat of the animal onto a dark, smooth surface.

3. Bunch the parasites together and transfer to a glass jar containing alcohol.

4. Slowly pull away the tick to avoid leaving the mouthparts in the skin or the animal. Place the removed tick in a glass jar containing alcohol. A dissecting scope can be used to identify the engorged nymph or adult tick.

5. To examine under a microscope, remove a few specimens from the alcohol with finely pointed forceps and place in a drop of water or oil on the microscope slide. Place a cover slip on top, being careful not to push down too hard. Morphological details of most species of lice and fleas can be seen with low magnification of a compound light microscope. A dissecting scope must be used for larger lice. Nymph and adult ticks that are attached and engorged can be removed from the skin with forceps.

6. The identification of lice and fleas as to genus and species is difficult and requires an entomologist specialized in the subject. In this exercise, we are just distinguishing between lice, fleas, and ticks.

Observations

Sketch your observations in the boxes below, include a drawing of a louse, a flea, and a tick. If any of these parasites were not found during the removal process, find an image of it on the internet.

SECTION FIVE

Plant Sciences

EXERCISE 17: Putting Down Some Roots—Root Anatomy

Materials Needed

✓ Monocot root slide
✓ Dicot root slide
✓ Compound microscope
✓ Set of colored pencils

Purpose

To study the structures and functions of parts of monocot and dicot roots.

Procedure

1. Study the functions of the monocot root parts described here, and identify them on the labeled diagram of a root cross-section.

   a. **Epidermis**—the outer layer of a monocot root; functions to protect the root from loss of water and from plant infections.

   b. **Mechanical tissue (hypodermis)**—a layer of thick-walled cells; serves as support and protection for the root.

   c. **Cortical parenchyma**—a thick layer of root cells lying between the hypodermis and the xylem.

   d. **Xylem**—composed of tubes of varying size that carry water absorbed by the root hairs from the roots to the other parts of the plant.

   e. **Phloem**—the small, closely packed cells found between the xylem and the pith; they carry food produced by photosynthesis in the leaf to the roots. The phloem is composed of sieve tubes, which are elongated, square-cornered cells of varying size and with perforated ends.

   f. **Pith**—the region of root cell maturation; composed of undifferentiated parenchyma cells, which are large and not closely packed.

   g. **Root tip**—the very short portion of the extreme free end of a root, where growth, water absorption, and the differentiation of the primary tissues take place.

   h. **Root cap**—thimble-shaped mass of loose cells that protect the growing point of the root.

   i. **Root hairs**—long, fingerlike structures projecting between the tiny particles of soil; absorb water for the entire plant.

2. Place the monocot root cross-section on the microscope stage centered over the light hole and fasten with the stage clips. Identify the root parts under the microscope.

3. Label the parts of the monocot root on the diagram in Observation 1 and color the parts as follows:

   a. epidermis—light green
   b. xylem—red
   c. phloem—yellow
   d. mechanical tissue—brown
   e. cortical parenchyma—orange

4. Study the functions of the dicot root parts described here, and identify them on the labeled diagram of a root cross-section.

   a. **Epidermis**—the outer layer of a dicot root; functions to protect the root from loss of water and from plant infections.
   b. **Parenchyma**—a thick layer of root cells lying between the epidermis and the endodermis.
   c. **Phloem**—the outer layer of the stele; serves as a food transport system.
   d. **Cambium**—cylinder of thin-walled cells that gives rise to new xylem to the inside and phloem to the outside. The second layer of the stele found between the phloem and the xylem.
   e. **Xylem**—tissue that carries water from the roots to the other parts of the plant.
   f. **Pericycle**—sheath of small, parenchymatous cells that surrounds the xylem and phloem.

5. Place the dicot root cross-section on the microscope stage centered over the light hole and fasten with the stage clips. Identify the root parts under the microscope.

6. Label the parts of the dicot root on the diagram in Observation 2, and color the parts as follows:

   a. epidermis—yellow
   b. parenchyma—orange
   c. endodermis—light green
   d. phloem—dark green
   e. cambium—red
   f. xylem—brown
   g. pericycle—blue

Observations

1. Label the parts of the monocot root on the following diagram and color the parts as outlined in Procedure 3.

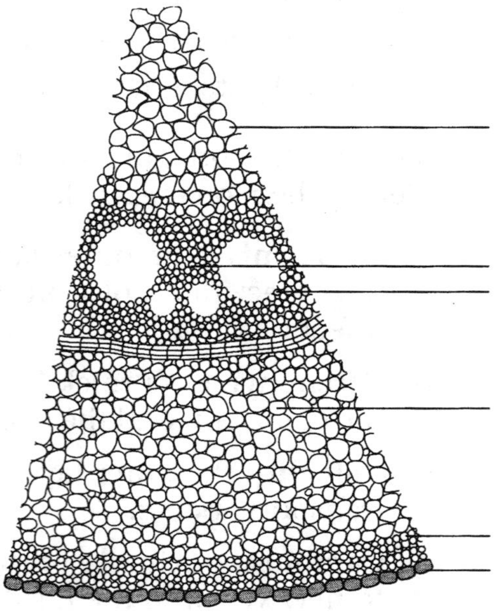

2. Label the parts of the dicot root on the following diagram and color the parts as outlined in Procedure 6.

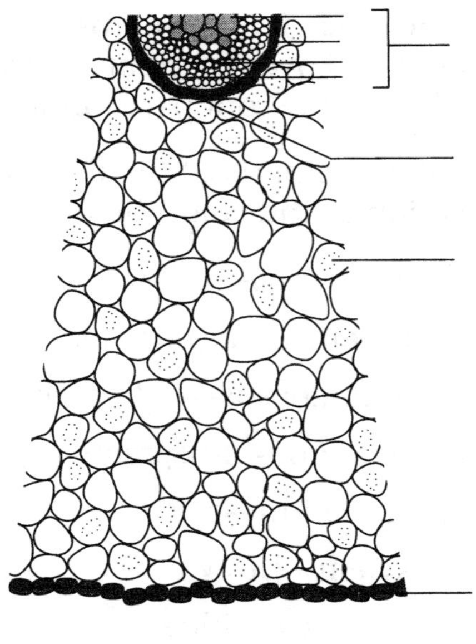

Conclusions

1. What composes most of the cells in a monocot root and in a dicot root?

____

____

____

2. What are the functions of the following in a plant root?

   a. xylem ______________________________

   ____

   b. phloem ______________________________

   ____

   c. cambium ______________________________

   ____

   d. epidermis ______________________________

   ____

   e. root cap ______________________________

   ____

   f. root hairs ______________________________

   ____

3. What functions do roots serve?

____

____

____

____

____

____

____

EXERCISE 18: It Stems from—Stem Anatomy

Materials Needed

- ✓ Monocot stem slide
- ✓ Dicot stem slide
- ✓ Compound microscope
- ✓ Set of colored pencils

Purpose

To study the structures and functions of the parts of monocot and dicot stems.

Procedure

1. Study the functions of the monocot stem parts described here, and identify them on the labeled diagram of a stem cross-section.
   a. **Epidermis**—the hard, green, outer layer of a monocot stem; functions to protect the stem from loss of water and from plant infections.
   b. **Fibro vascular bundle**—collectively, fibrous strands found in the monocot stem and composed of xylem, phloem, and mechanical tissue.
   c. **Xylem**—the inner part of the vascular bundle; carries water and nutrients from the roots to the other parts of the plant.
   d. **Phloem**—the small cells on the outer part of the vascular bundle; they carry food produced by photosynthesis in the leaf to other parts of the plant. The phloem is composed of sieve tubes, which are elongated, square-cornered cells of varying sizes and with perforated ends.
   e. **Mechanical tissue**—a layer of thick-walled cells surrounding the vascular bundle; provides support and protection.
   f. **Parenchyma**—cells that make up the pith of a monocot plant.
   g. **Air space**—air spaces found in the pith of the monocot stem.
2. Place the monocot stem cross-section on the microscope stage centered over the light hole and fasten with the stage clips. Identify the stem parts under the microscope.
3. Label the parts of the monocot stem on the diagram in Observation 1 and color the parts as follows:
   a. epidermis—light green
   b. xylem—red
   c. phloem—yellow
   d. mechanical tissue—brown

e. air space—white
f. parenchyma—orange

4. Study the functions of the dicot stem parts described in the following and identify them on the labeled diagram of a stem cross-section.

   a. **Cork**—outer layer of the bark; serves to protect the stem of the plant from disease and loss of water.
   b. **Cork cambium**—inner layer of the bark; gives rise to cells that become impregnated with a waxy substance, forming a waterproof layer that becomes cork.
   c. **Cortex**—cells that are found inside the cork cambium and serve as a food storage area.
   d. **Phloem**—the layer of cells that is found inside the cortex and serves as a food transport system. The phloem consists of phloem rays and phloem fiber.
   e. **Cambium**—cylinder of thin-walled cells that gives rise to new xylem to the inside and phloem to the outside.
   f. **Xylem**—tissue that carries water and forms the wood of a tree. The growth rings of a tree are found here.
   g. **Pith**—the center of a dicot stem; provides support to the plant during the first year of growth.
   h. **Vascular rays (phloem rays** and **xylem rays)**—horizontal strips of thin-walled cells; provide for the transverse conduction of water and nutrients along the tissues of the stem.

5. Place the dicot stem cross-section on the microscope stage centered over the light hole and fasten with the stage clips. Identify the stem parts under the microscope.

6. Label the parts of the dicot stem on the diagram in Observation 2 and color the parts as follows:

   a. cork—yellow
   b. cork cambium—orange
   c. cortex—light green
   d. phloem—dark green
   e. cambium—red
   f. xylem—brown
   g. pith—blue

Observations

1. Label the parts of the monocot stem on the following diagram and color the parts as outlined in Procedure 3.

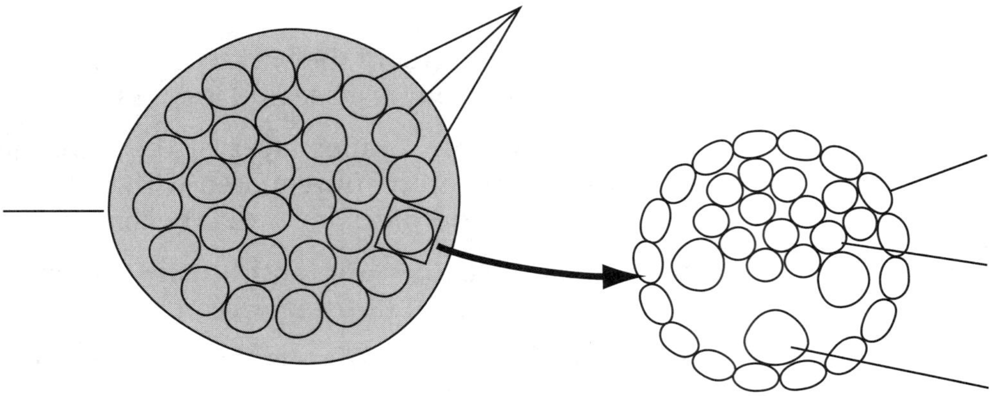

2. Label the parts of the dicot stem on the following diagram and color the parts as outlined in Procedure 6.

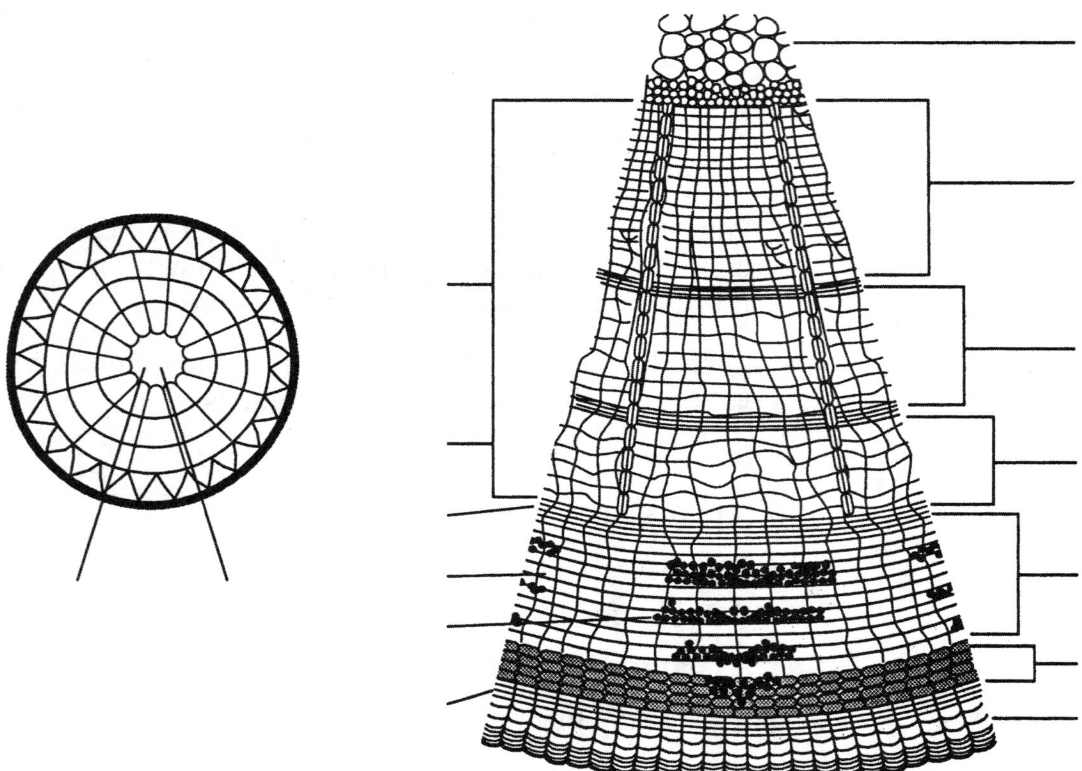

Conclusions

1. What composes most of the cells in a monocot stem and in a dicot stem?

2. What are the functions of the following in a plant stem?

   a. xylem

   b. phloem

   c. cambium

   d. epidermis

3. How can you determine the age of a dicot? How old is the one that you observed under the microscope?

4. What functions do stems serve?

5. If you girdle a tree (i.e., cut a ring around the tree through the bark), why does the tree die?

EXERCISE 19: Turn Over a New Leaf—Leaf Anatomy

Materials Needed

✓ Leaf section slide

✓ Compound microscope

✓ Set of colored pencils

Purpose

To study the structures and functions of parts of a typical plant leaf.

Procedure

1. Study the functions of the leaf parts described here, and identify them on the labeled diagram of a leaf cross-section.
   a. **Cuticle**—a very thin, non-cellular layer of waxy material found on both the top and bottom of the leaf.
   b. **Upper epidermis**—a single layer of heavy, walled cuboidal cells that serve as protection for the top part of the leaf.
   c. **Lower epidermis**—a single layer of heavy, walled cuboidal cells that serve as protection for the lower part of the leaf.
   d. **Chloroplasts**—the bodies in a leaf that contain chlorophyll, which imparts the green color to leaves.
   e. **Mesophyll**—the leaf tissue between the upper and lower epidermis; contains palisade cells, air space, and a spongy layer.
   f. **Palisade cell**—long, rectangular cells composing a layer one or two cells in thickness just below the upper epidermis; chloroplasts are found in these cells.
   g. **Air spaces**—spaces of air found between the spongy layer cells.
   h. **Spongy layer**—layer of loosely packed cells that is several cell layers thick; chloroplasts are found in this layer.
   i. **Stoma (Stomata)**—openings found in the lower epidermis of a leaf and through which moisture and gases pass.
   j. **Guard cells**—crescent-shaped cells that stand out from other epidermal cells because of their thickened walls and because they contain chloroplasts. They control the moisture content of the leaf and the exchange of gases.
   k. **Vascular bundle**—conducts water to the cells of the leaf and carries manufactured food from the leaf to the other parts of the plant. It is really a vein of the leaf itself.
   l. **Xylem**—thick-walled, upper-conducting cells in the vascular bundle.
   m. **Phloem**—thin-walled, lower-conducting cells in the vascular bundle.

2. Examine a leaf under a microscope.
   a. Place a prepared slide of a leaf cross-section on the stage. Center it over the light opening and fasten it in place with the stage clips.
   b. Focus with the low-power lens and observe the leaf parts. You may want to refer to the diagram in Observation 1.
   c. Label the leaf parts on the diagram in Observation 1.
   d. Find a vein of the leaf and switch to high power.
   e. Observe the xylem and phloem.
3. Color the leaf parts in the diagram in Observation 1 as follows:
   a. cuticle—orange
   b. upper epidermis and lower epidermis—yellow
   c. palisade cells—pink
   d. air space—white
   e. spongy layer—dark green
   f. stoma—light green
   g. guard cells—red

Observations

1. Review the definitions in one of the Procedures and refer to Figure 15-17 in *Agriscience: Fundamentals & Applications, 5th Edition.* Label the parts of the leaf in the following diagram, and color the parts as outlined in Procedure 3.

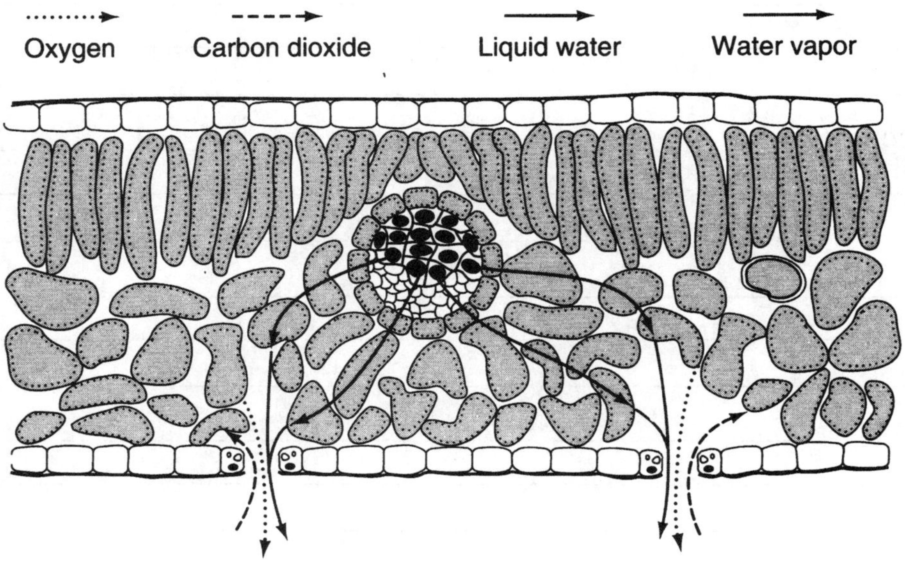

Conclusions

1. What is the main function of the leaf in a plant?

2. What function do chloroplasts serve?

3. Can you name some plants whose leaves would not contain chlorophyll?

4. What is the function of each of the following leaf parts?
   a. epidermis
   b. stomata
   c. guard cells
   d. cuticle
   e. chloroplasts

EXERCISE 20: He Loves Me, He Loves Me Not—Flower Anatomy

Materials Needed

- ✓ Taraxacum young bud slide
- ✓ Taraxacum older bud slide
- ✓ Mature anther slide
- ✓ Pollen germination slide
- ✓ Compound microscope
- ✓ Set of colored pencils

Purpose

To become familiar with the parts of a plant flower and the functions of the parts.

Procedure

1. Study the following parts and functions while observing the labeled diagram.

a. **Stamen**—collectively, the male reproductive parts of a plant flower: the filament, anther, and pollen.

b. **Filament**—long stalk that supports the anther. The anther is attached to the end of the filament.

c. **Anther**—the male sex organ of a plant flower; manufactures the pollen.

d. **Pollen**—the male reproductive cell of a plant flower.

e. **Pistil**—collectively, the female reproductive parts of a plant flower: the stigma, style, ovary, and ovules.

f. **Stigma**—the center part of the plant flower; receives the pollen from the anther.

g. **Style**—the pollen enters the stigma and travels down the style to the ovary.

h. **Ovary**—the female sex organ of the plant flower. Ovules are produced here.

i. **Ovules**—the female reproductive cells of a plant; they are fertilized by the pollen and eventually ripen into seeds.

j. **Perfect flower**—has all of the parts mentioned in 1a through 1i.

k. **Imperfect flower**—at least one but possibly more of the parts mentioned in 1a through 1i are missing.

l. **Petals**—parts of a flower that produce a scent that acts as an attractant to insects and other pollinators.

m. **Sepals**—serve as protection for the developing flower.

2. Place the young bud slide under the microscope and view it carefully. Draw the young bud in Observation 1.

3. Place the older bud slide under the microscope and view it carefully. Compare it to the young bud slide. Record any differences in Observation 2.
4. Place the mature anther slide under the microscope and view it carefully. Record what you see in Observation 3.
5. View the pollen germination slide. Make a drawing of your observation in Observation 4.
6. Complete the diagram in Observation 5.

Observations

1. Draw the young bud as seen under the microscope.

2. What differences were observed between the young bud and the older bud?

____

____

____

____

____

____

____

3. Draw the mature anther as seen under the microscope.

4. Draw pollen germination as viewed under the microscope.

5. Label the parts of the following diagram.

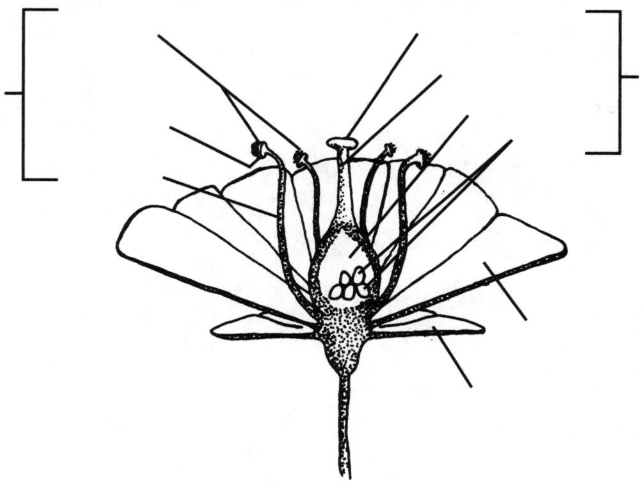

Conclusions

1. Identify the functions of the following flower parts.
   a. filament ______
   b. anther ______
   c. stigma ______
   d. ovary ______
   e. petals ______
   f. sepals ______

2. Explain the difference between a perfect and an imperfect flower. ______

3. Color the parts of the flower diagram in Observation 5 as follows:
   a. anther—green
   b. filament—light blue
   c. pollen —pink
   d. stigma—purple
   e. style—brown
   f. ovary—orange
   g. ovules—dark blue
   h. petals—yellow
   i. sepals—red

EXERCISE 21: Under the Knife—Flower Dissection

Materials Needed

- ✓ A complete flower
- ✓ Dissecting probe (teasing needles)
- ✓ Scissors
- ✓ Scalpel (optional)
- ✓ Tweezers/forceps (optional)
- ✓ Dissecting scope or zoom lens 18 to 108 mm

The development of fruit and seed begins with the flower for angiosperms. A complete flower differs from a perfect flower in that the complete flower contains all the floral whorls, whereas a perfect flower needs only to have a stamen and pistil.

Procedure

1. Pull the corolla from the flower petal-by-petal and record in the table below the number of petals.
2. Pull the calyx from the flower sepal-by-sepal and record in the table below the number of sepals.
3. Observe the stamens and pistil. Record the location of the pistil in the table below. Then record in the table below the location of the stamens.
4. Dissect the remaining parts of the flower lengthwise down the middle.
5. Pull out one stamen and sketch the parts of the stamen (2 parts). Make your sketch in the labeled box below.
6. Place a drop of water onto a microscope slide.
7. Holding a stamen with the tweezers or forceps over the drop of water on the microscope slide, gently tap the head of the stamen (the anther) with the dissecting probe. Pollen will fall from the anther to the drop of water. Examine the pollen using either a dissecting scope or zoom lens.

Observation

1. Sketch a few pollen grains in the labeled box below.
2. Pull out the pistil and draw the parts of the pistil (3 parts). Make your sketch in the labeled box below.
3. In the labeled box below, sketch the cup-shaped ovary that holds the ovules.

4. Sketch each part of the flower in the corresponding box.

Dissection of a Complete Flower	
Number of petals	
Number of sepals	
Location of pistil	
Location of stamen	

Anther	Filament	Pollen grain

Stigma	Style	Ovary

Ovules

Conclusions

1. What would make this flower an incomplete flower?

____

____

____

2. What would make this flower an imperfect flower? ______________________

____

____

EXERCISE 22: Free Gas Made Here—Photosynthesis

Materials Needed

- ✓ Calomba plant or other water plant
- ✓ Manometer
- ✓ Water
- ✓ Large beakers
- ✓ High-intensity light source
- ✓ Light-tight box
- ✓ Wax pencil

Purpose

One of the by-products of plant photosynthesis is the production of oxygen. A manometer is an instrument that measures gas pressure. For this lab, you may use the older style as shown here, or a newer digital one. In this exercise, you will be able to see how light intensity affects the amount of oxygen a plant will produce.

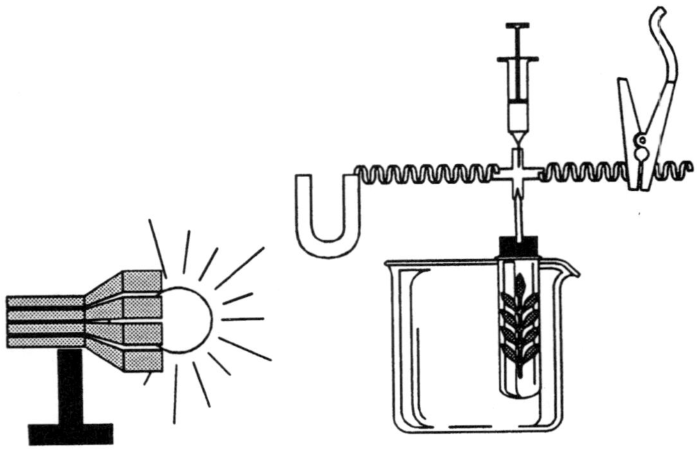

Procedure

1. Place the calomba (or other water plant) in a test tube with water, and seal the test tube with a stopper.
2. Complete the oxygen-measuring apparatus (manometer) as shown in the preceding diagram.
3. Place the test tube in the glass container of water, as illustrated in the preceding diagram.
4. Place the light approximately 2 feet from the apparatus on the side farthest away from the test tube.
5. Adjust the syringe as necessary to even the height of the colored fluid in the U-shaped tube to mark the position of the fluid. Use a wax pencil.
6. Let the light shine on the plant for 10 minutes and measure the height of the fluid. Record your findings in Observation 1.
7. Reposition the light 15 feet from the apparatus and use the syringes to readjust the height of the colored fluid in the U-shaped tube until the fluid is again even.

8. Let the light shine on the plant for 10 minutes and measure the height of the fluid. Record your findings in Observation 2.

9. Use the syringe to readjust the height of the level of the colored fluid to even one more time. Place a light-proof box over the apparatus.

10. Wait 10 minutes and measure the height of the fluid. Record your findings in Observation 3.

Observations

1. Direction of fluid movement in relation to test tube
   Amount of movement (in centimeters)

2. Direction of fluid movement in relation to test tube
   Amount of movement (in centimeters)

3. Direction of fluid movement in relation to test tube
   Amount of movement (in centimeters)

Conclusions

1. What caused the movement of the fluid in the manometer?

2. How did the light intensity affect the rate of photosynthesis?

3. Was any oxygen produced when the plant was in the dark? Why or why not?

4. What purpose did the beaker of water serve in this experiment?

____

____

____

____

EXERCISE 23: Transpiration—Nature's Big Gulp

Materials Needed

- ✓ 2 geranium leaves
- ✓ 2 small jars (baby food or Mason jars, or 2 beakers)
- ✓ Cardboard (10 cm square with a hole about the size of a hole-punch in center)
- ✓ Clay or petroleum jelly

Purpose

Transpiration is the process by which water is pulled through a plant, starting at the roots into the body of the plant and then evaporating into the atmosphere. In this exercise, you will be able to observe this process.

Procedure

1. Place a geranium leaf in the hole of the cardboard square. (Roll the geranium leaf to fit through the hole.) Seal the hole with clay or Silly Putty or petroleum jelly.
2. Fill one small jar with water.
3. Place the cardboard with the leaf over the jar of water. Make sure most of the leaf is in the water, but not so much that the cardboard comes into contact with the water.
4. Cover the leaf with a second jar. Place the leaf in a well-lighted area.
5. Observe the inside of the top jar after 24 hours.

Observation

1. Sketch your set up after you have completed it.
2. Sketch the set up again after 24 hours.
3. What changes took place?

Start	24 hours

Conclusions

1. Define equilibrium. ______________________________

____

____

2. What is osmosis? ______________________________

____

____

____

____

3. How is this experiment related to osmotic force or turgor? ______________________________

____

____

____

EXERCISE 24: Reap What You Sow—Seed Anatomy

Materials Needed

- ✓ Seed of a plant slide
- ✓ Compound microscope
- ✓ Several fresh bean seeds
- ✓ Magnifying lens
- ✓ Scalpel

Purpose

To become familiar with the parts and functions of the parts of seeds.

Procedure

1. Study the following parts of a seed:
   a. **Fruit coat**—transparent outer layer of the seed.
   b. **Endosperm**—makes up most of a seed and is composed of an opaque part, which contains starch, and a transparent part, which is composed of protein. These are the reserve foods that will be used by the embryo when it begins to grow.
   c. **Seed coats**—layers that surround the endosperm and serve to protect the seed's internal parts.
   d. **Embryo**—Externally, the part of the seed marked by a slight depression; internally, that part of the seed that grows into a new plant. Composed of the plumule, the plumule sheath, the cotyledon, the radicle, and the radicle sheath.
   e. **Cotyledon**—makes up most of the embryo. The cotyledon's infolded edges almost enclose the plumule. Collectively, the cotyledons are the seed leaves.
   f. **Plumule**—the primary bud of the seed embryo.
   g. **Plumule sheath**—made up of several immature leaves, which will become the first leaves of the plant.
   h. **Radicle**—becomes the root system of the plant.
   i. **Radicle sheath**—sheath of protection that surrounds the radicle.
   j. **Point of attachment**—where the seed is attached to the mature plant.
2. Place the slide of the corn seed under the microscope and identify the parts. Label these parts on the diagram in Observation 1.
3. Use the scalpel to cut one bean seed as shown in this diagram. Record what you find inside in Observation 2.

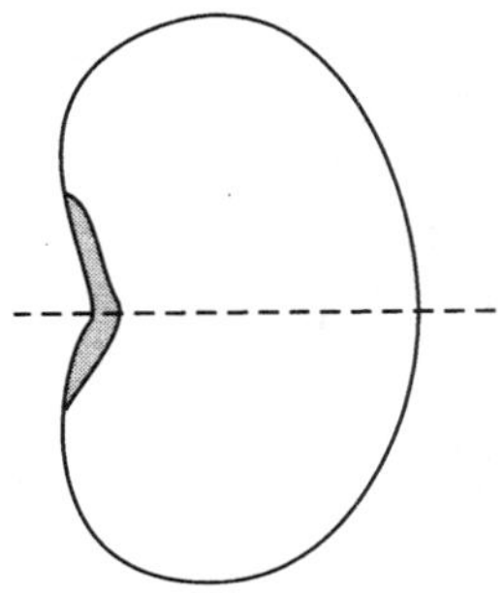

4. Cut another bean seed as shown here. Record what you see in Observation 3.

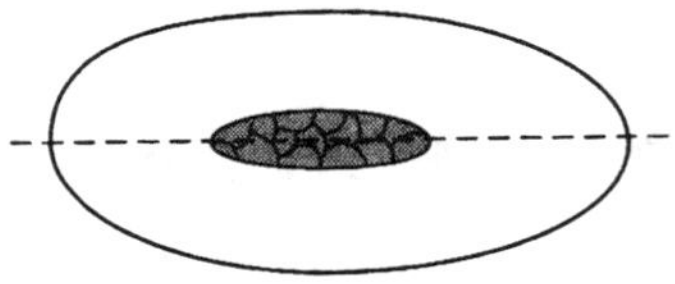

Observations

1. Label the parts of the seed in this diagram.

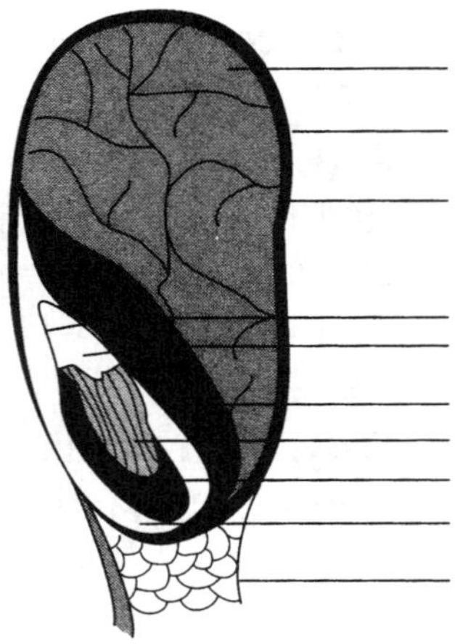

2. Draw what you observed inside the bean seed cut in Procedure 3.

3. Draw what you observed inside the bean seed cut in Procedure 4.

Conclusions

1. What is the function of the largest part of a seed?

____

____

____

____

____

____

____

2. What are the functions of the following seed parts?

   a. plumule ____

   ____

   b. radicle ____

   ____

   c. embryo ____

   ____

   d. seed coat ____

   ____

3. What is the function of seeds in the life cycle of a plant?

____

____

____

____

____

____

____

EXERCISE 25: Come Out, Come Out Wherever You Are—Germination

Materials Needed

- ✓ Water
- ✓ Aluminum foil
- ✓ Beaker
- ✓ Hot plate
- ✓ Freezer
- ✓ Seed-Gro Kit

Purpose

To determine how differences in environmental conditions affect seed germination.

Procedure

1. Moisture experiment
   a. Place a piece of dry aluminum foil in a dry place.
   b. Place five seeds on the foil or paper. *Do not water.*
   c. Place one of the foam germination pads in the petri dish and water as directed by your instructor.
   d. Place three seeds of the same kind as you placed in the foil on the foam pad, and add a few drops of water directly on the seeds to hasten germination.
   e. Replace the cover on the germination chamber to reduce moisture loss and to keep out unwanted materials.
   f. Observe the seeds for a 7-day period. Record your observations on the chart in Observation 1.
2. Air (oxygen) experiment
   a. Fill a beaker approximately half full with water.
   b. Place five seeds in the beaker.
   c. Place one of the foam germination pads in the petri dish and water as directed by your instructor.
   d. Place three seeds of the same kind as you placed in the beaker on the foam pad, and add a few drops of water directly on the seeds to hasten germination.
   e. Replace the cover on the germination chamber to reduce moisture loss and keep out unwanted materials.
   f. Observe the seeds for a 7-day period. Record your observations on the chart in Observation 2.
3. Temperature experiments
   a. Prepare two packets, each containing five seeds wrapped in aluminum foil.
   b. Place one packet in the freezer for one week.
   c. Place the other packet in boiling water for 10 minutes.

d. Prepare three germination chambers by placing the foam germination pads in the petri dishes and watering as directed by your instructor.

e. Label the germination chambers "hot," "cold," and "normal."

f. Plant the seeds boiled in water in the chamber labeled "hot," the freezer-conditioned seeds in the chamber labeled "cold," and the five untreated seeds in the chamber labeled "normal."

g. Observe the seeds for a 7-day period.

h. Record your observations on the chart in Observation 3.

4. Light experiment

a. Prepare two germination chambers by placing the foam germination pads in the petri dishes and watering as directed by your instructor.

b. Plant three seeds of the same kind in each chamber.

c. Wrap one chamber in aluminum foil, making sure no light can get into the chamber.

d. Observe the seeds every day for one week. Try to let no light (or as little as possible) get to the seeds in the aluminum-foil-wrapped chamber when you observe them.

e. Record your findings on the chart in Observation 4.

Observations

1. Record your moisture observations in the following chart:

	Day 1	Day 2	Day 3	Day 4	Day 5	Day 6	Day 7
Germination Chamber Seeds							
Foil-Wrapped Seeds							

2. Record the results of the air experiment in the following chart:

	Day 1	Day 2	Day 3	Day 4	Day 5	Day 6	Day 7
Germination Chamber Seeds							
Seeds in Beaker of Water							

3. Record the results of the temperature experiment in the following chart:

	Day 1	Day 2	Day 3	Day 4	Day 5	Day 6	Day 7
Normal Seeds							
Cold Seeds							
Hot Seeds							

4. Record the results of the light experiment in the following chart:

	Day 1	Day 2	Day 3	Day 4	Day 5	Day 6	Day 7
Germination Chamber Seeds							
Foil-Wrapped Seeds							

Conclusions

1. What part does moisture play in seed germination?

2. What happened to the seeds that were placed in the beaker of water? What caused this?

3. What effect did heat have on the germination of the seeds?

4. What effect did cold temperature have on the germination of the seeds?

5. How did the absence of light affect the germination of the seeds?

6. What conditions are necessary for proper seed germination?

EXERCISE 26: Let's See What You're Made of—DNA Extraction

Materials Needed

(per student pair):

- ✓ 1 plastic quart-size bag
- ✓ 1 strawberry
- ✓ 10 mL DNA extraction buffer
- ✓ Filter paper
- ✓ Ice cold 90 percent ethanol or Isopropyl alcohol
- ✓ 1 Test tube or champagne flute
- ✓ Funnel (optional)
- ✓ Inoculation loop or coffee stir straw
- ✓ Microscope
- ✓ Slides and Coversheets

Purpose

This lab will allow you to extract DNA and actually see what millions of strands of DNA look like.

Procedure

1. Place one thawed strawberry in the plastic bag and squeeze until it is a uniform puree.
2. Add 10 mL of buffer solution. Zip the bag closed.
3. "Moosh" the mixture of strawberry and buffer solution (without creating bubbles).
4. Fold filter paper or paper towel into a half circle, then a quarter circle, and open it to form a cone.
5. Fill the test tube (about 2 inches) or champagne flute to just above the stem with ice-cold alcohol. Place the filter paper cone into the test tube or champagne flute so that half of the cone is on the inside of the tube and half on the outside of the tube.
6. Fill the paper towel cone with the strawberry solution.
7. As the strawberry mixture filters through the cone and comes in contact with the alcohol, the DNA will form ribbons and then coagulate at the top of the alcohol.
8. Use the straw to spool the DNA.
9. Prepare a wet mount slide and observe the DNA under a microscope.

Observations

1. Did the DNA strands go to the bottom of the alcohol and then grow upward, or did the strands begin forming as soon as the strawberry solution made contact with the alcohol?

____

____

2. What color are the strands of DNA? ______________________

3. Did your group extract more or less DNA strands than other groups in the class?

____

____

4. Why do you think your group extracted more or less DNA strands?

____

____

5. In the boxes below, sketch your observations of the DNA without magnification and with magnification.

Without Magnification	With Magnification

Conclusions

1. If you tried to see a single thread from across the room, you probably would not be able to see it. But you would be able to see a thick rope made up of many thousands of threads twisted together. How is this similar to the DNA isolation you just performed?

____

____

____

____

____

2. How could you design this experiment to show that only living (or once living) organisms have DNA?

____

____

____

____

____

SECTION SIX

Crop Science

EXERCISE 27: We Are Here, We Are Here, We Are Here—Soil Organisms

Materials Needed

- ✓ What's in the Soil II—Microorganism Kit
- ✓ Compound microscope—300X or better
- ✓ Four 400-mL beakers
- ✓ Test tube rack
- ✓ Two test tubes
- ✓ Bunsen burner
- ✓ Handful of green leaves
- ✓ Soil samples
- ✓ Water
- ✓ Personal Safety Set

SAFETY NOTE

Always wear eye protection glasses and rubber gloves when handling specimens. Wash hands before leaving the laboratory.

Purpose

To recognize the different organisms found in the soil that aid in decomposition of organic matter and the formation of humus.

Procedure

1. Study the information on soil organisms provided in the microorganism kit.
2. Collect a soil sample to be used in this experiment. Record the information regarding the collection site in Observation 1.
3. Label one test tube "broth." Add nutrient broth to the test tube, filling the tube approximately two-thirds full. Keep the tube aside.
4. Label one test tube "agar." Pour the agar into the test tube and set the tube at a slant to allow the agar to harden. (Hardening at a slant provides a greater surface for organism growth.)
5. Place soil from your soil sample into a cup and add a small amount of water. Swirl the cup gently to mix the soil and water. Add more water so that the cup is two-thirds full.
6. Use a clean eyedropper and add a drop of the soil–water mixture to the tube containing the nutrient broth. Seal the tube and swirl it gently to mix the contents. Place the test tube in the rack. Check the tube in 24 hours. Record your results in Observation 2.
7. Sprinkle a small amount of soil on the agar in the slanted test tube. Seal the tube and place the test tube in the rack. Check the test tube in 48 hours. Record the results in Observation 3.
8. Wash a small piece of green leaf in water. Rinse the leaf in alcohol to remove any organisms from its surface and rinse again in water.

   WARNING: Alcohol is flammable. Do not expose to heat or flame.

9. Add the leaf to the cup containing the soil–water and place the cup in a location that receives sunlight. Be sure that you

have labeled the cup with your name. Check the cup in 48 hours. Record your findings in Observation 4.

10. Place a drop of water from the nutrient broth mixture on a glass slide, add a drop of methylene blue, and place a coverslip over the slide. Place the slide on the microscope and view first under low power and then under high power. Record your findings in Observation 5.

11. Place a piece of the growth from the agar solution on a slide, add a drop of methylene blue, and place a coverslip over the slide. Place the slide on the microscope and view first under low power and then under high power. Record your findings in Observation 6.

12. With an eyedropper, remove a small amount of water from near the surface of the leaf in the cup of soil. Place a drop on a slide and cover with a coverslip. View the slide under the microscope and record your findings in Observation 7.

Observations

1. Record the following information regarding the soil sample site:
   a. location of sample ______________________________
   b. light conditions ______________________________
   c. temperature conditions ______________________________
   d. humidity conditions ______________________________

2. Record your findings regarding the nutrient broth.

____

____

____

____

3. Record your findings regarding the agar test tube.

____

____

____

____

4. Record your observations regarding the soil water.

5. Draw or describe what you see on the slide of the nutrient broth and methylene blue.

6. Draw or describe what you see on the slide of the agar and methylene blue.

7. Draw or describe what you see on the slide of the soil water.

Conclusions

1. What soil organisms were observed in this experiment?

2. Why are these organisms found near the soil surface?

3. What functions do these organisms serve in the soil?

4. Were any of the organisms that you found plants? How can you tell?

EXERCISE 28: A Good Start—Growing and Transplanting Vegetable Seedlings

Materials Needed

- ✓ Tomato seeds
- ✓ Seed-starting media
- ✓ Flats
- ✓ Spray bottle with water
- ✓ Trowel or spoon
- ✓ Plastic wrap
- ✓ Large containers (10 ounces each or larger)
- ✓ Plant-growing media
- ✓ Liquid fertilizer
- ✓ Ruler
- ✓ Toothpicks

Purpose

To practice the proper methods of sprouting and transplanting seedlings.

Procedure

1. Study the following terms.

a. **flat**—a wooden or a plastic box with slotted bottom used to start seedlings

b. **germinate**—to sprout or begin to grow

c. **media**—materials that are used to start and grow seeds and plants

d. **seedlings**—young plants that have been germinated several days

e. **transplant**—to move plants from one growing location to another thus giving them more space in which to develop

2. Growing the seedlings

a. Fill the sections of the flat with the seed-starting media almost to the top.

b. Place something under the flat to catch any excess water.

c. Use the spray bottle to mist the mixture until it is thoroughly damp.

d. Place two tomato seeds in each compartment of the flat or in your assigned compartments if sharing a flat with a classmate.

e. Use the trowel or spoon and sprinkle additional seed-starting media on top of the seeds no more than $^1/_8$ inch deep.

f. Mist the mixture again.

g. Wait 10 to 15 minutes and mist once more with the spray bottle.

h. Place four toothpicks in a square on the outer edges of the flat. Place another toothpick in the center. Cover the flat with plastic wrap being careful not to cover it too tightly. The toothpicks will work as tent poles to hold the wrap off of the media.

i. Place the flat in a nice warm spot. Check daily for moisture. Mist if necessary.

j. Once the seeds sprout, they need light, either artificial or sunlight, for growth.

k. Once the tomatoes have grown their first set of true leaves, they are ready to be transplanted to larger containers.

3. Transplanting the seedlings

a. Place about 8 ounces of growing media in each container. (Containers should have drainage holes.)

b. Use the trowel or spoon to gently pry up the tomato seedlings.

c. Gently hold the seedling by its leaves (not stems) and carefully place it in its new container.

d. Firm the soil around it with your fingers.

e. Use the spray bottle to thoroughly moisten the soil in the container.

f. Make sure the containers are set where they receive adequate light and keep the soil moist (not wet).

g. After a week, provide the plants with a diluted (one-fourth to one-half strength) mixture of liquid fertilizer.

h. Plants may be transplanted to their final growing location in about 1 month.

Observations

1. Record the progress of your seed growth in the chart that follows. Record height and number.

Tomato Seeds' Progress	Day 2	Day 4	Day 6	Day 8	Day 10	Day 12
Number Sprouting						
Height of Tallest						
Height of Shortest						

2. Record the progress of your transplanted tomato plants for three weeks.

Tomato Plants' Progress	Day 2	Day 4	Day 6	Day 7	Day 14	Day 21
Height of Tallest						
Height of Shortest						

Conclusions

1. What were the important environmental conditions necessary for the sprouting of the seedlings?

____

____

____

____

2. What were the important environmental conditions necessary for the growth of the seedlings after sprouting?

____

____

____

____

3. What was your germination rate? ***Hint:*** Number of seedlings ÷ Number of seeds planted × 100.

____

____

____

____

4. If the germination rate was not 100 percent, give possible reasons.

____

____

____

____

5. Did all plants that were transplanted survive? Why or why not?

EXERCISE 29: Try That Oon For Size—Grafting Fruit Trees

Materials Needed

- ✓ Knife
- ✓ Grafting wax
- ✓ Grafting tape
- ✓ Apple or other fruit stock—¼-inch to ½-inch diameter
- ✓ Scions—¼-inch to ½-inch diameter

Purpose

To become familiar with the whip and tongue method of grafting fruit trees.

Procedure

1. Study the following terms related to grafting.

   a. **grafting**—uniting two different plants so that they grow as one

   b. **cambium**—thin, green, actively growing tissue located between the bark and wood of a plant

   c. **grafting wax**—a pliable, sticky, waterproof material made of beeswax, resin, and tallow

   d. **rootstock**—the lower portion of a graft that becomes the stem and roots of the new plant

   e. **scion**—a short piece of shoot containing several buds that becomes the new top of a grafted plant.

2. Making the cuts

   a. Be sure that the rootstock and scion are as close as possible to the same size.

   b. Using the knife, make the first cut on the scion below a bud. The length of the cut should be such that a smooth slanted angle is cut from 1½ to 2 inches long.

   c. The rootstock should be cut next with the same length and slope as that of the scion so that the two parts will fit together evenly.

   d. On the scion make a second cut beginning about one-third the distance from the tip. The cut should begin vertically but gradually become nearly parallel to the first cut surface. This is the tongue.

   e. Follow the same procedure as in the previous step with the rootstock. When finished, the scion and rootstock should fit together with the tongues interlocking.

3. Completing the graft
   a. Make sure that the cambium layers of the rootstock and scion are matched together on at least one side.
   b. Push the scion and rootstock together tightly matching the surfaces as closely as possible. Be sure there is no overlap of scion and rootstock.
   c. Wrap the graft with the grafting tape to keep it tight to prevent drying.
   d. Apply grafting wax over the entire area that is wrapped with tape as uniformly as possible. This is an additional precaution to prevent drying.
   e. In both the taping and the waxing steps, be careful not to dislodge the aligned cambium layers of the scion and rootstock.

Observations

1. In grafting the scion and rootstock together, why was it important to match the cambium layers together?

   ____

   ____

   ____

   ____

2. What could be the possible consequences of not wrapping and waxing the graft?

   ____

   ____

   ____

   ____

Conclusions

1. Define the following terms:
   a. cambium ____

      ____

   b. grafting ____

      ____

c. rootstock ____

____

d. scion ____

____

2. Why is grafting used in the fruit and nut industries?

____

____

____

3. Is grafting an economical method of producing fruits and nut trees? Explain your answer.

____

____

____

____

EXERCISE 30: From Fries to Fuel—Making Biodiesel

Materials Needed

The complete materials list is provided in the instructor's manual.

- ✓ 1 liter *new* vegetable oil
- ✓ 4 grams substance A (sodium hydroxide or caustic lye)
- ✓ 250 mL substance B (is methanol or wood alcohol)
- ✓ 2 glass containers with lids (glass juice bottle)
- ✓ 1 funnel
- ✓ 4 qt pot
- ✓ Warming plate
- ✓ Thermometer
- ✓ Blender (Do not use with food again)
- ✓ Turkey baster (Do not use with food again)
- ✓ Protective eye wear
- ✓ Thick rubber gloves

Purpose

With fuel costs at an all time high and exploration of new sources caught up in red tape, producers of biodiesel are finding their product in high demand. In this exercise, you will observe the process by which vegetable oil is used to produce fuel that can run your car, truck, or tractor. Due to chemical exposure, this exercise will be performed by your instructor.

Procedure

Note: Do not attempt to make biodiesel at home! Wear safety gloves and eye protection glasses during this exercise, and make sure that all items of equipment are clean and dry.

1. On the warming plate, heat the oil to 55° C in the 4 qt pot and remove it from the heat.
2. Pour the 250 mL of substance B into the glass container and quickly replace the lid. Be careful not to breathe in any fumes.
3. Measure out four grams of substance A. Be sure not to let this substance come into contact with any moisture. Using a clean and dry funnel, add substance A to substance B in the glass container. Replace the lid.
4. Keeping container away from your face, swirl the mixture gently for 1 minute. Check to see if the granules are dissolved. If they are not, swirl again for 1 minute. Repeat this process until all of substance A has completely dissolved (this can take up to six times). The mixture will become very warm as the chemicals combine.
5. Check the blender to make sure that all of the gaskets are in good condition and all parts are tightly fitted. Pour oil into the blender.
6. With the blender turned **OFF**, slowly add the A/B solution. Secure the blender lid and turn the blender on to a low speed. Allow the solution to mix for 20 minutes.
7. Transfer the solution to a clean glass jar. Within 20 minutes, you will begin to see a separation. After 24 to 36 hours, their will be two distinct layers. A clear yellow layer (biodiesel) and a thick brown layer (glycerin byproduct).

Observations

1. Describe the appearance of the three substances prior to mixing.

2. In your opinion, why do you think mixing substances A and B together produced heat?

3. Describe the appearance of the mixture soon after if was taken out of the blender.

4. Draw a picture of the mixture 20 minutes after it was taken from the blender. Label your picture.

Conclusions

1. Did the mixture produce two distinct layers after it had time to settle? If not, what do you think happened?

2. List three benefits of producing biodiesel.

3. As a farmer, what crop/s could you produce for use in the biodiesel industry?

EXERCISE 31: Deep Thoughts—Exploring Seed Planting Depths

Materials Needed

- ✓ Six glass jars—pint size
- ✓ Soil
- ✓ Water
- ✓ Corn seeds

Purpose

To observe the effects of planting depth on seed germination and crop production.

Procedure

1. Prepare the jars.

a. Label the jars "1" through "6." Prepare the jars as follows:

1. Jar 1—place 1 inch of loosely packed soil in the bottom of the jar and place one corn seed on each of the four sides of the jar next to the side where they can be seen; put 3 inches of loosely packed soil on top of the corn seeds.
2. Jar 2—place 2 inches of loosely packed soil in the bottom of the jar and place one corn seed on each of the four sides of the jar next to the side where they can be seen; put 2 inches of loosely packed soil on top of the corn seeds.
3. Jar 3—place 2½ inches of loosely packed soil in the bottom of the jar and place one corn seed on each of the four sides of the jar next to the side where they can be seen; put 1½ inches of loosely packed soil on top of the corn seeds.
4. Jar 4—place 3 inches of loosely packed soil in the bottom of the jar and place one corn seed on each of the four sides of the jar next to the side where they can be seen; put 1 inch of loosely packed soil on top of the corn seeds.
5. Jar 5—place 3½ inches of loosely packed soil in the bottom of the jar and place one corn seed on each of the four sides of the jar next to the side where they can be seen; put ½ inch of loosely packed soil on top of the corn seeds.
6. Jar 6—place 4 inches of loosely packed soil in the bottom of the jar and place one corn seed on each of the four sides of the jar next to the side where they can be seen; do not cover.

2. Moisten the soil thoroughly but not excessively.
3. Place the six jars where they will receive adequate light and warmth.
4. Monitor the progress of the seeds for three weeks, watering when needed.
5. Record your findings on the chart in Observation 1.

Observations

1. Record the progress of your seeds in the following chart.

Jar Number	Week 1	Week 2	Week 3
1			
2			
3			
4			
5			
6			

Conclusions

1. Which planting depth (or depths) was/were most appropriate for the corn seed? Why?

2. What happened to the seeds that were planted the deepest?

3. Why is it important to crop production for seeds to be planted at the proper depth?

4. If planted in a field, what would probably have happened to the seeds in jar 6?

5. What conditions are necessary for proper seed germination?

EXERCISE 32: Does This Smell Good?—Storing Silage

Materials Needed

- ✓ Freshly cut chopped grass
- ✓ Six 1-quart jars with tight-fitting lids
- ✓ Labels for the jars
- ✓ Pen
- ✓ Nails
- ✓ Hammer

Purpose

To observe how silage is produced and how different factors affect silage quality.

Procedure

1. Prepare the forage

a. Spread the grass out to dry for three days.

b. Label the jars "1" through "6."

c. Ensile the grass as follows:

1. Jar 1—pack the grass tightly, fill with water, and cover the jar tightly.
2. Jar 2—pack the grass tightly, fill with water, cover the jar, and punch several holes in the lid.
3. Jar 3—pack the grass tightly, cover the jar tightly.
4. Jar 4—pack the grass tightly, cover the jar, and punch several holes in the lid.
5. Jar 5—pack the grass loosely, cover the jar.
6. Jar 6—pack the grass loosely, do not cover the jar.

2. Storing the silage

a. Put the jars in a dark place.

b. Check them every day for 3 weeks to observe the development of the silage.

c. Compare the silage in your jars to the quality factors listed in the following table.

d. Record your results in Observation 1.

Comparing Silage Color and Odor			
Color		***Odor***	
Natural green color. Some clovers may turn a darker color than normal.	Good	Clean, pleasant, vinegary or pickled smell.	Good
Light green or yellowish color; also may be slightly brown.	Fair	Slightly fruity, yeasty, or scorched smell. Corn with high moisture may have acidic smell.	Fair
Gray or whitish color (molding); black or dark brown (overheating).	Poor	Musty smell (molding); strong burnt odor (overheating); rotten odor (poor bacterial action).	Poor

Observations

1. Complete the following chart regarding the quality characteristics of the silage in your jars.

Jar Number	Color	Odor	Other Observations
1			
2			
3			
4			
5			
6			

Conclusions

1. Which jar produced the best silage based on color and odor?

2. What happens when forage is ensiled too wet?

3. Why do you think silage is packed tightly when stored?

EXERCISE 33: Stop Buggin' Me—Insect Collection

Materials Needed

- ✓ Hand lens 10–30X power, or a dissecting scope
- ✓ Sweep net (can be purchased from any biological educational catalog)
- ✓ Stapler
- ✓ Paper bags (large grocery store size)
- ✓ Glass jars with screw-top lids
- ✓ Ethanol or isopropyl alcohol (rubbing alcohol)
- ✓ Use of a freezer (optional)
- ✓ Glass dish or Petri dish
- ✓ Internet access

Purpose

To identify the types of insects that can cause harm to a field crop.

Sweep Method: Importance to Industry

The use of a sweep net is a very common method of monitoring insects in a field. In crops such as alfalfa and cotton, insects are collected and counted to estimate the population levels. Field monitoring is important to determine whether treatment is necessary.

In this exercise, the use of a sweep net is demonstrated in alfalfa. Insects of concern are the alfalfa weevil larvae (*Hypera postica*) or Egyptian alfalfa weevil (*Hypera brunneipennis*), alfalfa caterpillars (*Colias eurytheme*), beet armyworm (*Spodoptera exigua*), and western yellow-striped armyworm (*Spodoptera praefica*). Other insects, including predators and parasitic wasps (which are the good guys in this case) can also be detected.

Procedure

1. A standard sweep net consists of a cone-shaped cloth bag approximately 2 feet deep fitted to a 15-inch wire loop and attached to a 26-inch handle. To sweep, tilt the rim of the net so that the lower edge is 1 to 2 inches ahead of the upper edge of the rim. This allows the net to catch insects falling from the plant. The lower rim of the net should be kept about 10-inches below the top of the plant during sweeping. A single sweep is on a 180-degree arc taken during a step forward. At the end of this sweep, raise the net and reverse the direction of the swing. Swing the body from side to side during sweeping, making natural swings. Sweeps may be taken singly or consecutively.
2. To check a field, divide the field into four or more sections depending on the size of the field. Take five sweeps in each section and deposit the contents of the net into a paper bag and staple the bag closed. Place the paper bags from each section of the field into a freezer for 10 minutes. Open the bags and discard the plant material. Pour the insects into the glass jars and add enough alcohol to cover the insects.

. Seal the jars tightly to prevent the alcohol from evaporating. The alcohol preserves the samples for counting and identification at a later time.

3. To count and identify the insects, carefully pour off as much alcohol as possible without losing any insects. Place the insects in a shallow glass dish or Petri dish. If the insects float, there is too much alcohol remaining in the sample. Put the dish under the microscope and examine the sample. Identify the different insects and count the number of each species. To determine the average number per sweep, add the number of each species from each section of the field and divide by the total number of sweeps taken from the field.

Observations

	Average Number per Sweep			
Insect:				
Number per Sweep				
Number of Sweeps				
Average Number per Sweep				

- The recommended treatment threshold for weevils is an average of 20 larvae per sweep.
- Caterpillars have a number of predators, parasites, and pathogens that may occur in a field that will keep the populations below economically damaging levels. The treatment threshold for alfalfa caterpillars is an average count of 10 non-parasitized or disease-free caterpillars per sweep.
- Beet armyworm or western yellow-stripe armyworm need to be controlled when an average of 15 non-parasitized armyworms of more than ½ inch in length per sweep are found.

Control of these insects could be obtained with early harvest of the hay or with an insecticide. When economically practical, many growers choose to harvest early to minimize the killing of predators and parasites of aphids.

Conclusions

1. Using the information above and the data you collected, do you believe that a treatment plan should be established for this field? Why?

2. Using the internet answer the following, what are the current practices used to control one of the insects you found?

SECTION SEVEN

Ornamental Use of Plants

EXERCISE 34: Keep 'em in the Dark—Plants and Light

Materials Needed

- ✓ Six pots of pea seedlings
- ✓ Four light-tight boxes, each with a hole in one side
- ✓ One completely light-tight box
- ✓ Light source

Purpose

To determine how plants respond to different lighting conditions.

Procedure

1. Water the plants well so that you will not need to disturb them during this experiment.
2. Arrange the four light-tight boxes so that their holes face different directions. For example, one box has its hole toward the east side, one has its hole toward the west side, one has its hole toward the north side, and one has its hole toward the south side.
3. Place a light source (a lamp will work) in the middle of the four boxes.
4. Place a pot of pea seedlings in each box.
5. Place one pot of pea seedlings where it gets light from all directions.
6. Place one pot of pea seedlings in the completely light-tight box.
7. Leave the plants in the boxes for one week. At the end of the week, open the boxes and observe the plants.
8. Record the plants' responses to light on the chart in Observation 1.

Observations

1. Complete the following chart regarding the plants' response to light.

Lighting Condition	Plant Response		
	Color of Plant	Direction of Growth	Length of Stem
Light from North			
Light from South			
Light from West			
Light from All Directions			
No Light			

Conclusions

1. What was the general effect of light versus darkness on the pea plants' growth? How can you explain this?

2. How did the color vary on the plants exposed to light versus the one kept in total darkness?

3. How did the plants move in response to the light?

EXERCISE 35: Chip Off the Old Block—Plant Cloning

Materials Needed

- ✓ Plant II Cloning: Propagation of African Violets Kit
- ✓ African violet plant
- ✓ Soap
- ✓ Bunsen burner
- ✓ Scalpel
- ✓ Water—Sterile
- ✓ Four to five large beakers
- ✓ African violet potting soil
- ✓ Ethyl alcohol
- ✓ Hot plate
- ✓ Laundry bleach
- ✓ Ten to twelve small plastic pots
- ✓ Personal Safety Set

SAFETY NOTE

Always wear eye protection glasses and rubber gloves when handling dissection specimens. Wash hands before leaving the laboratory.

Purpose

To observe the methods of cloning plants to obtain uniformity in production.

Procedure

1. Study the sterile-technique procedures outlined in the cloning kit instruction guide. Follow these procedures when working with this equipment.
2. Follow the procedure outlined in the kit for preparation of shoot multiplication media.
3. Prepare the leaves for culture as described in the kit instruction guide.
4. Establish the shoot multiplication culture as outlined in the guide. Observe the plants every 2 weeks for 3 months. Record your findings on the chart in Observation 1.
5. Prepare the rooting medium according to the directions in the kit.
6. Transfer the plants to the rooting medium as outlined in the instruction guide found in the kit. Observe the plants for the next three weeks. Record your findings in Observation 2.
7. After numerous roots have appeared, carefully transfer the plants to pots containing African violet potting soil. Scrape all excess agar from the roots. Treat the young plants as you would any other plant.

Observations

1. Record the plant progress in the following chart:

First 2 Weeks	Second 2 Weeks	Third 2 Weeks	Fourth 2 Weeks	Fifth 2 Weeks	Sixth 2 Weeks

2. Record your findings in the following chart:

First Week	Second Week	Third Week

Conclusions

1. What type of plant reproduction is cloning?

2. What advantages can you think of for cloning plants? What disadvantages?

3. Do you think that cloning is a very productive method of reproduction in animals? Why?

4. What are some of the current controversies surrounding animal cloning?

SECTION EIGHT

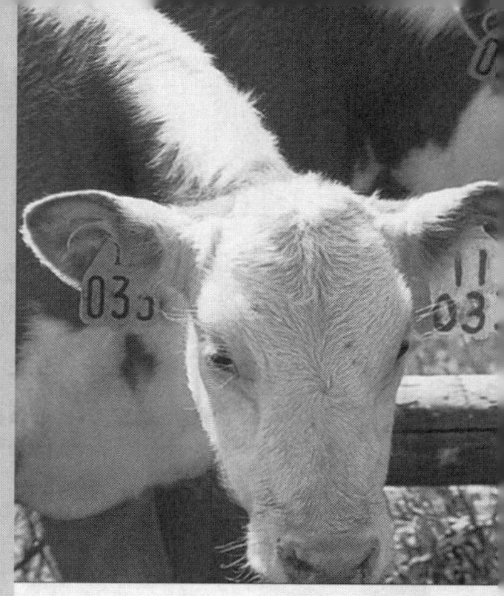

Animal Sciences

EXERCISE 36: You Eat Like a Pig—Simple Animal Digestion

Materials Needed

- ✓ Textbook
- ✓ Rubber gloves
- ✓ Scalpel
- ✓ Dissecting microscope
- ✓ Personal Safety Set
- ✓ Preserved digestive tract of a pig or a fetal pig

SAFETY NOTE

Always wear eye protection glasses and rubber gloves when handling dissection specimens. Wash hands before leaving the laboratory.

Purpose

Humans have a digestive system that is very similar to that of a pig. In this exercise, you will examine and identify the parts and functions of a simple digestive system.

Procedure

1. Study the following parts of the simple digestive system in Figure 26-9 in *Agriscience: Fundamentals & Applications, 5E* and locate them on the drawing.

Alimentary Canal

a. **Mouth**—the beginning of the simple digestive tract; external opening through which food is ingested.

b. **Pharynx**—passageway from the mouth to the esophagus. Muscles here help move the food into the esophagus.

c. **Esophagus**—muscular passageway from the pharynx to the stomach. A flap of skin at the entrance to the esophagus called the epiglottis prevents food from entering the trachea.

d. **Stomach**—gastric juices continue the breakdown of food in this part of the simple digestive system.

e. **Small intestine**—food is further broken down for absorption here, as pancreatic juices and bile are added to the digesting food.

f. **Caecum**—a blind gut; serves no function in swine but functions to help rabbits and horses digest forages.

g. **Large intestine**—most moisture is squeezed from the undigested food here; also called the colon.

h. **Rectum**—part of the alimentary canal where undigested food is held prior to release from the body.

i. **Anus**—external opening of the digestive tract through which undigested food is expelled from the body.

Accessory Organs

a. **Teeth**—serve to grind and pulverize food.

b. **Salivary glands**—secrete digestive salivary juices (saliva), which begin the chemical breakdown of food.

c. **Pancreas**—produces pancreatic digestive juices, which are released into the small intestine to help break down food.

d. **Liver**—lobe-shaped organ that purifies the blood brought from the stomach, pancreas, and intestines. It also secretes bile, which is later added to the digestive juices in the small intestine.

e. **Gallbladder**—stores some bile from the liver and empties bile into the large intestine as waste to be removed from the body.

2. Wearing rubber gloves and safety goggles, spread out the digestive tract so that all parts can be easily seen. Identify the following parts and label them on the diagram in Observation 1.

a. From the mouth, feed passes through the esophagus and into the stomach. Locate the esophagus.

b. Follow the esophagus to the stomach. Trace the stomach toward the posterior of the tract until it joins the small intestine. The small intestine is the smaller tube that leads to the large intestine (the larger tube).

c. Where the small intestine and large intestine join is the caecum, a blind gut.

d. Follow the large intestine toward the posterior. There is a slight enlargement just before it reaches the anus. This is the rectum.

e. Find the liver. It is a lobed, dark red-brown organ.

f. Look for the gallbladder. It is a small, greenish sac embedded under the right lobe of the liver.

g. The pancreas is a small, pinkish, grainy organ found inside the bend made by the first section of the small intestine.

3. Using a scalpel, remove a 2-inch section of the small intestine and cut it lengthwise. Wash its lining and examine the intestine section under a dissecting microscope. In Observation 2, draw what you see. The projections you see are called villi. They aid in the absorption of nutrients into the blood-stream from the small intestine.

Observations

1. Label the parts of the simple digestive system on the following diagram.

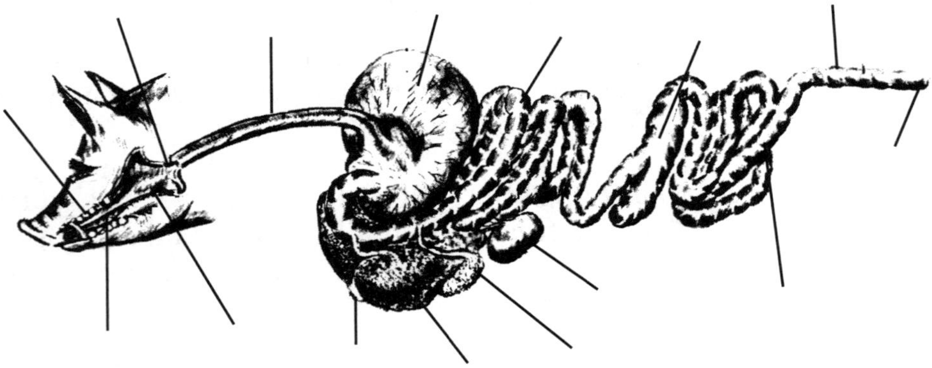

2. Draw the lining of a small intestine.

Conclusions

1. What is the function of the digestive system?

___

___

___

___

2. Identify the parts of the simple digestive system that function as follows:
   a. Squeezes moisture from the undigested food ___
   b. Produces bile ___
   c. Grinds and pulverizes the food ___
   d. Muscular passageway from the pharynx to the stomach ___
   e. Accessory organ that is found in the first bend of the small intestine and produces digestive juices to aid in digestion ___

3. Which part of the digestive system digests roughages in horses and rabbits?

___

___

___

___

4. What purpose do villi serve in the small intestine?

___

___

___

___

EXERCISE 37: It Takes Four—Ruminant Digestion

Materials Needed

- ✓ Textbook
- ✓ Rubber gloves
- ✓ Scalpel
- ✓ Four jars
- ✓ Dissecting microscope
- ✓ Compound microscope
- ✓ Rumen bacteria slide
- ✓ Personal Safety Set
- ✓ Preserved digestive tract of a cow or sheep

SAFETY NOTE

Always wear eye protection glasses and rubber gloves when handling dissection specimens. Wash hands before leaving the laboratory.

Purpose

To identify the parts and the functions of the parts of the ruminant digestive system.

Procedure

1. Study the following parts of the ruminant digestive system in Figure 26-8 of *Agriscience: Fundamentals & Applications, 5E* and locate them on the drawing.

Alimentary Canal

a. **Mouth**—the beginning of the simple digestive tract; external opening through which food is ingested.

b. **Pharynx**—passageway from the mouth to the esophagus. Muscles here help move the food into the esophagus.

c. **Esophagus**—muscular passageway from the pharynx to the stomach. A flap of skin at the entrance to the esophagus called the epiglottis prevents food from entering the trachea.

d. **Rumen**—the first compartment of a ruminant's stomach. It is sometimes called the paunch and makes up approximately 80 percent of the ruminant stomach. It acts as a large storage vat where food is agitated and fermented, and where digestion begins.

e. **Reticulum**—this part of the ruminant stomach has a honeycomb-like lining that can catch and hold foreign objects that may be picked up by the animal. For this reason, it is sometimes referred to as the hardware stomach. It composes approximately 5 percent of the ruminant stomach. It, along with the rumen, has many bacteria and protozoa that aid in the breakdown of forages (roughages).

f. **Omasum**—the third part of the ruminant stomach; grinds up food and may squeeze water from the food. It is sometimes called the manyplies and composes approximately 8 percent of the ruminant stomach.

g. **Abomasum**—this part of the ruminant stomach is sometimes called the true stomach. It works similarly to the stomach in the simple digestive tract in that digestive juices are produced here. The abomasum composes approximately 7 percent of the ruminant stomach.

h. **Small intestine**—food is further broken down for absorption here, as pancreatic juices and bile are added to the digesting food.

i. **Caecum**—a blind gut; serves no function in ruminants but functions to help rabbits and horses digest forages.

j. **Large intestine**—most moisture is squeezed from the undigested food here; also called the colon.

k. **Rectum**—part of the alimentary canal where undigested food is held prior to release from the body.

l. **Anus**—external opening of the digestive tract through which undigested food is expelled from the body.

Accessory Organs

a. **Teeth**—serve to grind and pulverize food.

b. **Salivary glands**—secrete digestive salivary juices (saliva), which begin the chemical breakdown of food.

c. **Pancreas**—produces pancreatic digestive juices, which are released into the small intestine to help break down food.

d. **Liver**—lobe-shaped organ that purifies the blood brought from the stomach, pancreas, and intestines. It also secretes bile, which is later added to the digestive juices in the small intestine.

e. **Gallbladder**—stores some bile from the liver and empties bile into the large intestine as waste to be removed from the body.

2. Wearing rubber gloves and safety goggles, spread out the digestive tract so that all parts can be easily seen. Identify the following parts and label them on the diagram in Observation 1.

a. From the mouth, feed passes through the esophagus and into the rumen. Locate the esophagus.

b. Follow the esophagus to the rumen. Trace the rumen to the reticulum. There is no real division between the rumen and reticulum.

c. Next locate the omasum. Follow it until you find the abomasum.

d. Follow the abomasum until it joins the small intestine. The small intestine is the smaller tube that leads to the large intestine (the larger tube).

e. Where the small intestine and large intestine join is the caecum, a blind gut.

f. Follow the large intestine toward the posterior. There is a slight enlargement just before it reaches the anus. This is the rectum.

3. Label the jars "rumen," "reticulum," "omasum," and "abomasum."

4. Cut the stomach open and take out some of the contents from each compartment. Put the contents in the jars labeled accordingly. Compare the digestion of food in each compartment. Record your findings in Observation 2.

5. Cut out a section of the inner lining of each part of the stomach. Wash each section thoroughly and observe it under the dissecting microscope. Draw what you see in Observation 3.

6. Place the rumen bacteria slide on the stage of the microscope and clip it down. Focus on low power and switch to high power. In Observation 4, draw the bacteria that you see.

Observations

1. Label the parts of the ruminant digestive system in the following diagram.

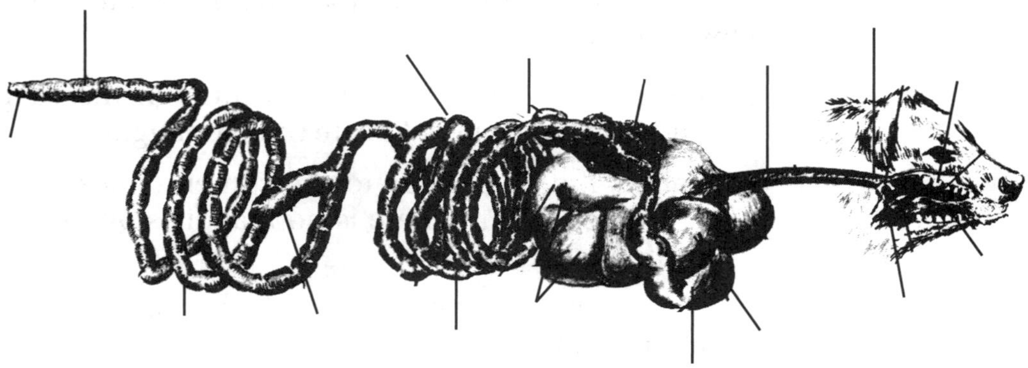

2. Describe the contents of each of the parts of a ruminant stomach.

a. rumen ____
____
____

b. reticulum ____
____
____

c. omasum ____
____
____

d. abomasum ____
____
____

3. Draw the inner lining of each part of the ruminant stomach as observed under the microscope.

a. rumen

b. reticulum

c. omasum

d. abomasums

4. Draw the bacteria as seen under the microscope.

Conclusions

1. Identify the function of each of the parts of the ruminant stomach.
   a. rumen ______
   b. reticulum ______
   c. omasum ______
   d. abomasum ______

2. What type of feed can ruminants digest that simple stomachs cannot?

______

3. How do bacteria help digestion in the ruminant stomach?

______

4. Which part of the ruminant stomach performs the same basic function as does the simple stomach?

______

5. Define the term "ruminant."

______

EXERCISE 38: Must Have Been Something I Ate—Internal Parasites

Materials Needed

- ✓ Sheep Liver Fluke Microscope Slide Set
- ✓ Sheep Tapeworm Microscope Slide Set
- ✓ Pig Roundworm Microscope Slide Set
- ✓ Compound microscope

Purpose

To study the structure and life cycles of several economically important internal parasites.

Procedure

1. Study the information sheets for this exercise before continuing.
2. Examine the sheep liver fluke (*Fasciola hepatica* life history) slides under the microscope. Draw the liver fluke eggs in Observation 1.
3. Examine the sheep liver fluke infected snail tissue under the microscope.
4. Examine the sheep liver fluke infected sheep liver under the microscope. Compare the two slides of infected tissue in Observation 2.
5. Examine the sheep tapeworm slides under the microscope. Draw the tapeworm eggs in Observation 3.
6. Examine the pig roundworm slides under the microscope. Draw the pig roundworm eggs in Observation 4.
7. Study the life cycle charts of the three internal parasites.

Observations

1. Draw the liver fluke eggs.

2. Compare the snail tissue damage to the liver damage. Draw or describe what you see.

3. Draw the tapeworm eggs.

4. Draw the pig roundworm eggs.

5. Complete the following life cycle charts by drawing arrows to indicate the cycle flow.

a.

b.

c.

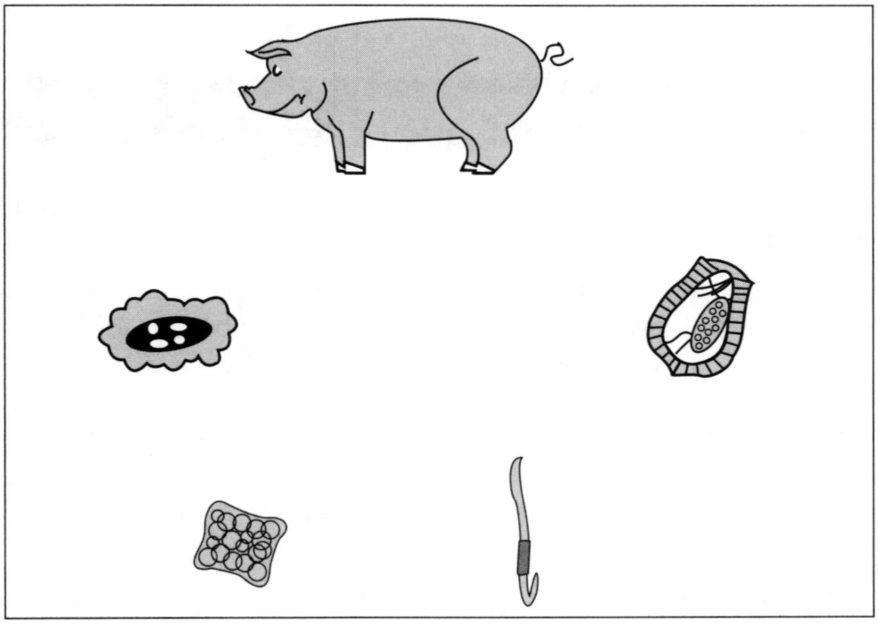

Conclusions

1. What type of damage does each of the following cause?
   a. liver fluke ______________________________
   b. tapeworm ______________________________
   c. roundworm ______________________________

2. How do these three internal parasites differ in the damage that they cause? How are they the same?

____

____

____

____

3. What differences are there in the life cycles of these parasites? What similarities are there?

____

____

____

4. How are most internal parasites controlled?

____

____

____

INFORMATION SHEET — LIVER FLUKE

The liver fluke is a parasite of cattle, sheep, goats, and humans. It is especially damaging to young animals.

The adult lives in the bile ducts, where eggs are laid and pass down into the intestines and out in the feces. The eggs must land in water to hatch. The larvae that hatch from these eggs swim about seeking a snail. The snail is necessary for completing the life cycle. The larvae develop for a period in the snail, then emerge and encyst on plants along the water. Livestock eat the water plants and become infected. The young flukes pass to the intestines and burrow through the abdominal cavity and into the liver, where they live principally on blood. Egg production begins approximately 3 months after entering the animal.

The fluke causes irritation, thickening of the bile duct, and fibrosis of the liver, making the animal unfit for human consumption.

The symptoms of liver flukes are anemia and weight loss. Highly infested animals may die.

Rotation of pastures and the use of water troughs help to control liver flukes. Chemical treatment will kill the adult flukes in the animal. Control of snails will break the cycle but is hard to do.

INFORMATION SHEET — TAPEWORM

There are several species of tapeworm that affect livestock. The broad tapeworm affects all classes of livestock and humans.

The adult lives in the small intestines, where it may reach a length of 10 or more feet. Segments of the tapeworm that contain eggs are continuously breaking off and passing out in the feces. The eggs are eaten by the orbatid mite, which lives in the grass and weeds and serves as an intermediate host. The eggs develop in these mites, then are eaten by livestock, and hatch in the small intestines. They feed on animal foodstuff and grow to maturity. There is no physical damage to the host; rather tapeworms compete with the host for food.

Symptoms of tapeworms include unthriftiness, loss of weight, diarrhea, and emaciation.

Tapeworms can be eliminated from their hosts via the use of chemicals.

INFORMATION SHEET — ROUNDWORM

Roundworms are the most important group of internal parasites from an economic standpoint. There are many types of roundworm, which affect almost every type of livestock. The roundworms of greatest concern are found in the digestive system, mostly in the stomach and intestines.

Roundworms in the adult stage live as blood-sucking parasites attached to the stomach wall. The eggs pass from the host to the feces and hatch into larvae in 15 to 20 days, depending on the temperature and humidity. The larvae crawl up blades of grass, are eaten by the animal, and travel to the stomach lining until they mature.

In the process of penetrating the stomach lining before maturing, they cause severe damage by reducing the digestion of nutrients by the host and producing poisons. Young, undernourished, or diseased animals are hardest hit.

The most common symptom is anemia, which can be indicated by a paleness of the gums and the whites of the eyes. In light infestations, the animal will have a dull hair coat, an unthrifty appearance, and will sometimes scour. In severe infestations there will be persistent scouring, weight loss, anemia, prostration, and sometimes death.

Sanitation and pasture rotation are good control measures. Chemical dewormers are used in treating infested animals.

EXERCISE 39: It's All in the Genes—Understanding Genetics

Materials Needed

- ✓ Pencil or pen
- ✓ Calculator
- ✓ Text
- ✓ Internet access (optional)

Purpose

To provide students with an understanding of how different gene combinations result in different individuals.

Procedures

1. Study the following terms related to genetics:
   a. **heredity**—the passing on of traits or characteristics from parents to offspring
   b. **genes**—components of cells that determine the individual characteristics of living things
   c. **genetics**—the study of heredity
   d. **genotype**—what the genes look like
   e. **phenotype**—physical appearance of an individual
   f. **dominant**—gene that expresses itself to the exclusion of other genes
   g. **recessive**—gene that expresses itself only in the absence of a dominant gene
   h. **heterozygous**—pairs of genes that are different
   i. **homozygous**—pairs of genes that are alike
   j. **incomplete dominance**—neither gene expresses itself to the exclusion of the other
2. Complete the squares for the mating of the following two animals: Bb—Black bull and bb—Red cow Bb × bb (Observation 1)
3. Complete the squares for the mating of the following two animals: PP—Polled bull and pp—Horned cow PP × pp (Observation 2)
4. Complete the squares for the mating of the following two animals: BbPP—Black polled bull and bbpp—Red horned cow BbPP × bbpp (Observation 3)
5. Complete the squares for the mating of the following two animals: RR—Red Shorthorn bull and WW—White Shorthorn cow RR × WW (Observation 4)

Observations

1. Complete the squares that follow:

C O W

B U L L

2. Complete the squares that follow:

C O W

B U L L

3. Complete the squares that follow:

C O W

B U L L

4. Complete the squares below:

C O W

B U L L

Conclusions

1. Using the text or the internet, define the following terms:
   a. dominant ______________________________
   ____
   b. gene ______________________________
   ____
   c. genetics ______________________________
   ____
   d. genotype ______________________________
   ____
   e. heredity ______________________________
   ____
   f. heterozygous ______________________________
   ____
   g. homozygous ______________________________
   ____
   h. incomplete dominance ______________________________
   ____
   i. phenotype ______________________________
   ____
   j. recessive ______________________________
   ____

2. In Observation 1, identify the following:
   a. The dominant characteristic ______________________________
   ____
   b. The recessive characteristic ______________________________
   ____
   c. The phenotypic ratio ______________________________
   ____
   d. The genotypic ratio ______________________________
   ____

3. In Observation 2, identify the following:
   a. The dominant characteristic ______________________________
   ____

b. The recessive characteristic ______________________________

____

c. The phenotypic ratio ______________________________

____

d. The genotypic ratio ______________________________

____

4. In Observation 3, identify the following:

a. The dominant characteristic ______________________________

____

b. The recessive characteristic ______________________________

____

c. The phenotypic ratio ______________________________

____

d. The genotypic ratio ______________________________

____

5. In Observation 4, identify the following:

a. The dominant characteristic ______________________________

____

b. The recessive characteristic ______________________________

____

c. The phenotypic ratio ______________________________

____

d. The genotypic ratio ______________________________

____

EXERCISE 40: This Will Hurt Only for a Minute—Injection Procedures

Materials Needed

- ✓ Two or three 10-mL syringes
- ✓ ½-inch needles
- ✓ 1-inch needles
- ✓ Three containers of water
- ✓ Food dye—red, green, blue
- ✓ Three or four grapefruit
- ✓ Ruler

Purpose

To practice procedures for properly vaccinating animals for disease prevention and control.

Procedures

1. Study the following terms related to injection procedures.
   a. **injection**—the process of administering drugs by needle and syringe
   b. **antibiotic**—substance used to help prevent or control certain diseases of animals
   c. **intradermal**—between layer of skin
   d. **intramuscular**—in a muscle
   e. **intraperitoneal**—in the abdominal cavity
   f. **intrarumenal**—in the rumen
   g. **intravenous**—in a vein
   h. **subcutaneous**—under the skin
   i. **syringe**—an instrument used to give injections of medicine or to draw body fluids from animals
   j. **vaccination**—the injection of an agent into an animal to prevent disease
2. Add blue dye to one water container, red dye to a second water container, and green dye to the third. Add dye to the water until the water is a dark color.
3. Using the ½-inch needle, inject one grapefruit with 2 mL of green-colored dye directly under the skin of the grapefruit (subcutaneous).
4. Using a 1-inch needle, inject the same grapefruit used in Step 3 with 2 mL of red-colored dye directly into the grapefruit (intramuscular).
5. Using the ½-inch needle, inject one grapefruit with 5 mL of blue-colored dye directly under the skin of the grapefruit (subcutaneous).

6. Using a 1-inch needle, inject the same grapefruit used in Step 5 with 5 mL of green-colored dye directly into the grapefruit (intramuscular).

7. Using the ½-inch needle, inject one grapefruit with 10 mL of red-colored dye directly under the skin of the grapefruit (subcutaneous).

8. Using a 1-inch needle, inject the same grapefruit used in Step 7 with 10 mL of blue-colored dye directly into the grapefruit (intramuscular).

9. Wait 10 minutes then cut each grapefruit along a line where the intramuscular injection was given. Record your findings in Observation 1.

10. Peel the grapefruit around the area where the subcutaneous injection was given. Record your findings in Observation 2.

Observations

1. Provide the information requested in the following for the intramuscular injection:

Intramuscular Injection	**Grapefruit 1 2 mL**	**Grapefruit 2 5 mL**	**Grapefruit 3 10 mL**
Approximate Size of Area Colored with Dye			

2. Provide the information requested in the following for the subcutaneous injection:

Subcutaneous Injection	**Grapefruit 1 2 mL**	**Grapefruit 2 5 mL**	**Grapefruit 3 10 mL**
Approximate Size of Area Colored with Dye			

Conclusions

1. Which type of injection distributed the dye to the largest area?

2. Was the amount of area covered with the dye proportional from the smallest to the largest? Give reasons why it was or was not.

3. Which type of injection was easier for you to perform? Why?

4. Besides intramuscular and subcutaneous, list other types of injections that may be given.

EXERCISE 41: We Are Doing What?—Fecal Examination for Parasites

Materials Needed

- ✓ Compound microscope
- ✓ Plastic bag (Ziploc® type)
- ✓ Jelly bean size of fresh feces of a cat, dog, sheep, cattle, hog, or horse
- ✓ Microscope slide and coverslip
- ✓ Small piece of newspaper
- ✓ Toothpick
- ✓ Tap water
- ✓ Eyedropper
- ✓ Gloves and eye protection

Importance to Industry

The internal parasites with the most economic impact on livestock and dogs and cats include the roundworms (nematodes), flukes (trematodes), and tapeworms (cestodes). Heartworm (*Dirofilaria immitis*) is the leading cause of death in dogs.

Procedure

Note: This exercise should be performed by the instructor as an observation lab.

For the Collection of the Fecal Sample

Fresh feces should be collected and used. At least several grams of fecal material should be gathered (about the size of a jelly bean).

1. Turn the plastic bag inside out.
2. Place your hand in the bag and pick up the fresh fecal material.
3. Pull the plastic bag over your hand so that the fecal sample is now inside the bag and your hand is outside the bag.
4. Zip the bag closed.

Smear Technique

1. Place a clean microscope slide on a small piece of newspaper.
2. Put one drop of tap water onto the center of the slide.
3. Using a toothpick, remove a small sample of the collected fecal material.
4. Mix the sample into the drop of water on the slide until the suspension is cloudy but the newsprint can be read.
5. Using the form below, sketch and label one item from each of the fecal samples observed.

FECAL SAMPLE MICROSCOPE EXAMINATION

Name________________________ Date ________________

Label ______ Species ______	Label ______ Species ______	Label ______ Species ______	Label ______ Species ______
Label ______ Species ______	Label ______ Species ______	Label ______ Species ______	Label ______ Species ______
Label ______ Species ______	Label ______ Species ______	Label ______ Species ______	Label ______ Species ______
Label ______ Species ______	Label ______ Species ______	Label ______ Species ______	Label ______ Species ______

6. Remove any large bits of debris that may be present by scraping the toothpick horizontally over the cloudy solution (more drops of water may be needed).
7. Gently place the coverslip onto the slide.
8. Examine under the microscope. Begin with 10X power and then increase the power to see detail. When worm eggs and coccidial oocytes or other protozoa are present in the same sample, the smaller coccidian and giardia will float to the top toward the coverslip. When the power is 40X or higher, the eggs will be in focus, but the coccidian will not. The fine-adjustment knob will focus all three types of parasites.

Conclusions

1. Which samples appeared to have evidence of parasites?

____

____

____

____

2. What treatment(s) would you provide these or this animal with?

____

____

____

____

EXERCISE 42: Hot Wings Anyone?—Animal Anatomy and Physiology

Materials Needed

- ✓ One raw chicken wing for each student (Chicken wings can be purchased at the local supermarket in the inexpensive family pack.)
- ✓ Scissors
- ✓ Paper plate
- ✓ Tweezers or forceps
- ✓ Protective gloves

Purpose

To practice manipulating dissecting equipment and to become familiar with organs and tissue found throughout the body of animals.

Introduction

A bird's wing is made up of groups of tissues and organs working together to perform a job. Before beginning the dissection, review the functions performed by the tissues and organs.

- Skeletal muscles are attached to bones.
- Skin is the membranous tissue that forms the outer covering of the body that provides a protective barrier from the outside environment.
- Muscle tissue is composed of bundles of skeletal muscle fibers. When these tissues expand and contract they produce motion in the wing.
- Fatty tissue, when stored on the underside of the skin, helps to keep the body warm, cushions and protects other body tissues, and stores vitamins A, D, E, and K.
- Blood vessels are the arteries, veins, and capillaries, which transport blood throughout the body. Capillaries are too small to be seen without a microscope. Arteries have thicker walls than veins.
- Tendons are especially strong connective tissue that attaches skeletal muscle to bones.
- Cartilage helps prevent neighboring bones from grinding against each other.
- Bone provides structural support and manufactures red blood cells.
- Nerves are the bundles of fibers that transmit sensory stimuli and motion impulses.
- Ligaments are strong bands of connective tissue that connect two bones together.

Procedure

Read the description of each of the tissues or organs, then begin the dissection for that particular tissue/organ. In the box provided, sketch an illustration of each tissue/organ listed in this lab.

Skin—The skin is the external covering of the entire wing. The skin will have a weblike appearance between the bones. Look for evidence that the skin was covered with feathers. Cut a slit in the skin covering the largest bone and joint.

Skin

Connective Tissue—First lift a corner of the skin and, with tweezers or forceps, peel it back gently from the muscle. Notice the shiny, thin membrane that surrounds the muscle and attaches the muscle to the skin. This is connective tissue.

Connective Tissue

Muscle Tissue—Note the pink-orange bundles of fibers attached to the bone. This is muscle tissue.

Muscle Tissue

Fatty Tissue—Continue to peel back the skin slowly and gently until you locate a tissue between the skin and muscle that is yellow in color and greasy to the touch. If no fatty tissue can be found there, find a thick piece of skin and proceed to cut through it to see whether there is any fatty tissue. If the lab is utilizing store-bought chicken wings, it is possible that little or no fat can be observed.

Fatty Tissue

Blood Vessels—With the skin peeled back, scan the surface of the skeletal muscles for thin red tubes. These are blood vessels. Arteries and veins might also be located in bundles of connective tissue with nerves.

Blood Vessels

Tendons—Look at the top of the large bone. Notice the shiny white pieces of connective tissue that attach skeletal muscle to bone. These are tendons.

Tendons

Cartilage—Examine the top of the large bone. Locate a pearly white hard tissue found at the ends of the long bones. This is cartilage.

Cartilage

Bone—Cut through the muscle with your scissors to expose the hard white bone. Bones are hard (but not solid) tissue made up primarily of calcium and phosphorus. The scissors will not be able to cross-section the bone. Ask the instructor to cut the bone. Look into the cross section of the bone; the mass of red tissue is red marrow. It is the red marrow that manufactures red blood cells. Notice the bone surrounding the red marrow. The bone itself is not solid, but is an asymmetrical lattice work of calcium and phosphorus.

Marrow

Cross Section of Bone

Nerves—Nerves are usually found buried deep within an organism, often lying close to long bones. Sometimes an artery, vein, and nerve are held together in one bundle of connective tissue. Gently peel the muscle from the long bone and search carefully for a thin, white, thread-like tissue. This is a nerve. If this does not work, look for a bundle of connective tissue containing blood vessels. Gently pry it apart and search for the nerve.

Nerves

Ligaments—Cut through the skin and muscle down to the joint between the long bone and the center bone. Locate the strong white bands of connective tissue that connect the two bones together at the joint. This is a ligament.

Ligaments

SECTION NINE

Food Science and Technology

EXERCISE 43: Break It on Down?—Food Nutrients

Materials Needed

- ✓ Twenty test tubes
- ✓ Test tube rack
- ✓ Five eyedroppers
- ✓ Hot plate
- ✓ Sudan IV
- ✓ Biuret reagent
- ✓ Potassium iodide—Iodine solution
- ✓ Benedict's solution
- ✓ Milk
- ✓ Cream or butter
- ✓ Syrup
- ✓ Canned whole corn
- ✓ Distilled water

Purpose

To observe the different food nutrients found in the foods we eat.

Procedure

1. Group the test tubes in the test tube rack in groups of four. Label the five different groups as follows:
   a. Glucose—sugar
   b. Starch
   c. Fat
   d. Protein
   e. Water
2. Put 5 mL of water into each of the test tubes.
3. Add the following to the test tubes:
   a. Three to five drops of syrup to the four test tubes marked "Glucose."
   b. Three to five drops of juice from the canned corn to the four test tubes marked "Starch."
   c. Three to five drops of cream or melted butter to the four test tubes marked "Fat."
   d. Three to five drops of milk to the four test tubes marked "Protein."
   e. Three to five drops of distilled water to the four test tubes marked "Water."
4. Add 10 drops of Benedict's solution to the first tube of each group of four test tubes. Heat the four tubes gently in a hot water bath for 10 minutes. If sugar is present, the color will change to orange. Record your findings in the table in Observation 1.
5. Add one drop of potassium iodide—iodine solution—to the second tube in each group of four test tubes. Starch turns blue in the presence of iodine. Glycogen (animal starch) turns mahogany red in the presence of iodine. Record your findings in the table in Observation 2.

6. Add 1 drop of Sudan IV to the third tube in each group of four test tubes. Sudan IV is an oil-soluble dye and will tag oil globules. Record your findings in the table in Observation 3.

7. Add 10 drops of Biuret reagent to the last test tube in each group of four test tubes. Biuret reagent changes from blue to violet pink if peptides (protein building blocks) are present. The color may fade. Allow 10 to 15 minutes for the reaction to take place. Record your findings in the table in Observation 4.

Observations

1. Sugar test results

Test Tube	Reaction (if any)
Syrup	
Corn juice	
Cream	
Milk	
Water	

2. Starch test results

Test Tube	Reaction (if any)
Syrup	
Corn juice	
Cream	
Milk	
Water	

3. Fat test results

Test Tube	Reaction (if any)
Syrup	
Corn juice	
Cream	
Milk	
Water	

4. Protein test results

Test Tube	Reaction (if any)
Syrup	
Corn juice	
Cream	
Milk	
Water	

Conclusions

1. Which foods reacted to the starch test?

____

____

____

____

____

2. Which foods showed the presence of sugar?

____

____

____

____

____

3. What were the protein foods?

4. Which foods reacted to the fat test?

5. What reactions took place in the distilled water?

6. List the foods (if any) that showed more than one of the food nutrients and which nutrients were present.

EXERCISE 44: What's for Dinner?—Exploring the Food Guide Pyramid

Materials Needed

- ✓ Variety of foods from each food group
- ✓ Kitchen measuring tools
- ✓ Scale in ounces
- ✓ Internet access

Purpose

Agriculture is the most important industry in the world. We could live even without our laptops and ipods, but not without food. In this exercise, we will discover the kinds and amounts of food one person need to live a healthy lifestyle.

Procedure

1. Log onto www.mypyramig.gov and select *My Pyramid Plan.*
2. Fill out your personal information or that of a 16 year old160lb male that is 5'10", and who is active 30–60 minutes a day.
3. Review the information provided and fill out the chart in question number one of the conclusions sections.
4. For each food category, read the tips the web site provides to help individuals fill their daily nutritional need.
5. Using the foods provided by your instructor, measure out 10 ounces of food that would fall into the grains food group. Use at least three different kinds of grain-based foods to equal the total number of ounces your body needs. Keep in mind that half of the grains you consume throughout the day should come from whole grains.
6. Again, using the food provided by your instructor, measure the appropriate amount of food you should eat daily from each of the remaining food groups. Be sure to use at least two different food sources to represent each category.

Conclusions

1. Fill out the chart below to reflect how much food from each category you should eat each day:

Food Group	Amount
Grains	
Vegetables	
Fruit	
Milk	
Meat/Beans	

2. For each food group, list two tips that will help you fill the daily recommendations.

3. What are whole grains? How can you tell if a food falls into this category?

4. What are the recommendations for oils and discretionary calories provided?

EXERCISE 45: Is This Still Good?—Food Preservation

Materials Needed

- ✓ Bread
- ✓ Swiss cheese
- ✓ Hot plate
- ✓ Two 20-mL test tubes
- ✓ Two 250-mL beakers
- ✓ Distilled water
- ✓ Test tube holder
- ✓ Sulfuric acid solution
- ✓ Funnel
- ✓ Filter paper
- ✓ Spatula
- ✓ Ethyl alcohol
- ✓ Pencil
- ✓ Personal Safety Set

Purpose

To determine the presence of food additives in the foods that we eat.

Procedure

1. Place 400 mL of distilled water into a 600-mL beaker and bring to a low boil on the hot plate.
2. Cut the Swiss cheese into small squares and place in a 250-mL beaker. Add 100 mL of distilled water.
3. Mash the cheese and stir with a spatula to mix the cheese and water as much as possible. Allow to sit while preparing the bread.
4. Place a slice of bread in a 250-mL beaker and add 100 mL of distilled water. Let the bread soak. Stir the mixture with a spatula to mix the bread and water as much as possible.
5. Using the funnel with filter paper, filter approximately 10 mL of the cheese–water mixture in one test tube.
6. Follow the same procedure for the bread mixture.
7. Add 5 mL of sulfuric acid solution to both the Swiss cheese and bread solutions. Record the reaction in Observation 1.
8. Add 3 mL of ethyl alcohol to each test tube and place in water bath on hot plate.
9. When the solutions begin to boil, carefully observe the odor coming from the vapors. In Observation 2, record the odor you detect.

Note: An odor similar to pineapple indicates Ethyl propionates, which are mold inhibitors.

Observations

1. What happened to the cheese solution when the sulfuric acid was added?

2. What odor did you detect when you heated the cheese solution? the bread solution?

Conclusions

1. Why are preservatives necessary in foods such as cheese and bread?

2. What are some natural preservatives found in food?

3. If food preservatives were not used, how do you think it would affect food supply and prices?

EXERCISE 46: How Cheesy—Dairy Product Production

Materials Needed

- ✓ 12-inch piece of cheesecloth
- ✓ Hot plate
- ✓ Plastic cups—16 ounces each
- ✓ Cooking thermometer
- ✓ Milk
- ✓ Salt
- ✓ Rennin or rennilase

Purpose

To demonstrate the process involved in producing cheese.

Procedure

1. Heat the milk to 88° to 90° F on the hot plate. Use the thermometer to monitor the temperature.
2. Fill the plastic cups approximately two-thirds to three-fourths full.
3. Immediately add 12 drops of the rennin solution to the milk. Stir continuously until the rennin enzyme is completely mixed in.
4. Let the milk set undisturbed until the curd (solid portion) has separated from the whey (liquid portion).
5. Filter the liquid from the solid by folding the cheesecloth in half and placing over the top of the cup to act as a strainer.
6. Pour the liquid into another plastic cup.
7. Salt the cheese (curds) and allow it to further harden.
8. Cut and eat.

Observations

1. Describe the color of the curds. Describe the color of the whey.

2. Describe the taste of the finished cheese.

Conclusions

1. What function does the enzyme rennin serve in this experiment?

2. How do you explain the different tastes of different cheeses?

EXERCISE 47: Does This Milk Make Me Look Fat?—Homogenization

Materials Needed

- ✓ Compound microscope with an oil immersion lens
- ✓ Microscope slide and coverslip
- ✓ Raw milk from a dairy or a health food store that carries raw milk
- ✓ Homogenized milk
- ✓ Internet access

Purpose

Before milk hits the shelves of your local grocery store, it is homogenized. In this process, the fat in the milk is broken up to prevent it from separating. In this exercise, you will be able to see the difference between homogenized milk and raw milk.

Procedure

1. Put a drop of raw milk on one slide and a drop of homogenized milk on another slide.
2. Place a coverslip over each.
3. With the oil immersion lens, look at the differences in the size of the fat cells.

Observations

Sketch the fat globules of the homogenized milk sample and the fat globules from the raw milk sample.

Homogenized fat globules	**Raw fat globules**

Conclusions

1. Describe the differences between the two milk samples.

___

___

___

2. Using the Internet or other reference materials, define homogenization.

___

___

___

3. After researching the process involved in homogenization, give a brief description of your findings.

___

___

___

EXERCISE 48: Do You Have a Sweet Tooth?—Fruit Sugars

Materials Needed

- ✓ Refractometer
- ✓ A variety of fruits
- ✓ Kitchen knife
- ✓ Blender (optional)

Purpose

To measure sugar levels in fruits.

Procedure

1. Obtain a drop of liquid (juice) from a piece of fruit. This can be done by squeezing or mixing it up in a blender. Open the plastic top of the refractometer and drop a bead of the juice of the fruit on the glass window.
2. Close the plastic top.
3. Hold the refractometer so that the light source (outside sun is best) is directed toward the plastic top, not the end of the refractometer.
4. Read the degree brix at the point where the white and blue colors meet.
5. Record the degree brix in the chart on the following page.
6. Repeat the steps above for all of the fruit samples.

Observations

Before measuring the actual brix, list the fruits that will be used in this lab in the left column. Then list the fruits in the order in which you *think* they rank in sweetness from highest to lowest.

Fruits	Ranking from Highest to Lowest in Sweetness

Percent Sugar or Degree Brix

Fruit	Brix	Comments

Conclusions

1. What does a refractometer actually measure?

____

2. What is the name of the sugar found in fruits?

____

3. Of your favorite fruits, which do you think is the sweetest?

____

SECTION TEN

Communications and Management in Agriscience

EXERCISE 49: Where Did the Money Go?—Understanding Diminishing Returns

Materials Needed

✓ Pencil

✓ Calculator

Purpose

To provide an understanding of the law of diminishing returns.

Procedure

1. Study the terms associated with diminishing returns:

   a. **marginal**—refers to incremental changes, increases or decreases, which occur at the edge or margin (additional)

   b. Δ—shorthand for "change in"

   c. **total physical product (TPP)**—the total output or yield

   d. **average physical product**—average amount of output produced by each unit of input at each input level:

   $$APP = \frac{\text{Total physical product (TPP)}}{\text{Input level}}$$

   e. **marginal physical product (MPP)**—the additional or extra total physical product produced by using an extra unit of input:

   $$MPP = \frac{\Delta \text{ total physical product}}{\Delta \text{ input level}}$$

   f. **marginal value of product (MVP)**—the additional or marginal income received from using an additional unit of input:

   $$MVP = \frac{\Delta \text{ total value of product (TVP)}}{\Delta \text{ input level}}$$

   g. **total value of product (TVP)**—the product of multiplying quantity of output (TPP) by selling price.

   h. **marginal input cost (MIC)**—the change in total input cost or the addition to total input cost caused by using an additional unit of input:

   $$MIC = \frac{\Delta \text{ total input cost}}{\Delta \text{ input level}}$$

i. The **profit maximizing point** is the input level where MVP = MIC.

2. Study the following examples:

Input price = $10.00 per unit
Output price = $2.50 per unit

Input Level	TPP	APP	MPP	TVP ($)	MVP ($)	MIC ($)
0	0	–	–	0	–	–
1	10	10	10	25	25	10
2	28	14	18	70	45	10
3	42	14	14	105	35	10
4	52	13	10	130	30	10
5	60	12	8	150	20	10
6	66	11	6	165	15	10
7	70	10	4	175	10	10
8	72	9	2	180	5	10
9	70	7.78	—2	175	—5	10
10	68	6.80	—4	170	—10	10

a. Input level and total physical product (TPP) are given. Input price and output price are given.

b. Study the **a**verage **p**hysical **p**roduct (APP) column. Examples:

$$APP = \frac{TPP}{input} = \frac{10}{1} = 10 \text{ for level 1}$$

$$\text{APP} = \frac{\text{TPP}}{\text{input}} = \frac{28}{2} = 14 \text{ for level 2}$$

c. Study the **m**arginal **p**hysical **p**roduct (MPP) column.
Examples:

$$\text{MPP} = \frac{\text{TPP}}{\Delta\ \text{input}} = \frac{(10-0)}{(1-0)} = 10 \text{ for level 1}$$

$$\text{MPP} = \frac{\text{TPP}}{\Delta\ \text{input}} = \frac{(28-10)}{(2-1)} = 18 \text{ for level 2}$$

d. Study the **t**otal **v**alue of **p**roduct (TVP) column.
Examples:

$$\text{TVP} = \text{TPP} \times \text{selling price} = 10 \times \$2.50 = \$25.00 \text{ for level 1}$$

$$\text{TVP} = \text{TPP} \times \text{selling price} = 28 \times \$2.50 = \$70.00 \text{ for level 2}$$

e. Study the **m**arginal **v**alue of **p**roduct (MVP) column.
Examples:

$$\text{MVP} = \frac{\text{TVP}}{\Delta\ \text{input}} = \frac{(25-0)}{(1-0)} = \$25.00 \text{ for level 1}$$

$$\text{MVP} = \frac{\text{TVP}}{\Delta\ \text{input}} = \frac{(70-25)}{(2-1)} = \$45.00 \text{ for level 2}$$

f. Study the **m**arginal **i**nput **c**ost (MIC) column.
Examples:

$$\text{MIC} = \frac{\text{total input cost}}{\Delta\ \text{input}} = \frac{(1-0)(\$10.00)}{(1-0)} = \frac{10}{1} = 10$$

$$\text{MIC} = \frac{\text{total input cost}}{\Delta\ \text{input}} = \frac{(2-1)(\$10.00)}{(2-1)} = \frac{10}{1} = 10$$

g. The profit maximizing point is where MVP = MIC. If you look at input level 7, you find MVP = 10 and MIC = 10.

h. The law of diminishing returns states that as additional units of a variable input are used in combination with one or more fixed inputs, the marginal physical product will eventually begin to decline. Study the MVP column on the chart.

Observations

Complete the following chart.

Fertilizer costs $15.00 per unit.

Corn sells for $3.00 per bushel.

Units of Fertilizer	TPP (Corn)	APP	MPP	TVP ($)	MVP ($)	MIC ($)
0	70	–	–	0	–	–
1	96					
2	118					
3	129					
4	136					
5	140					
6	150					
7	140					
8	136					

1. Complete the **a**verage **p**hysical **p**roduct (APP) column.
2. Complete the **m**arginal **p**hysical **p**roduct (MPP) column.
3. Complete the **t**otal **v**alue of **p**roduct (TVP) column.
4. Complete the **m**arginal **v**alue of **p**roduct (MVP) column.
5. Complete the **m**arginal **i**nput **c**ost (MIC) column.

Conclusions

1. What is the profit-maximizing amount of fertilizer to use on corn?

2. If the TVP column is higher on level 6, why is it not the profit-maximizing level?

3. If fertilizer cost only $12.00 per unit, what would be the profit-maximizing level of fertilizer to use? Why?

EXERCISE 50: I Owe You—Interest and Credit

Materials Needed

- ✓ Pencil
- ✓ Calculator

Purpose

To demonstrate the different methods of charging interest for borrowing money.

Procedure

1. Study the following terms:

   a. **principal**—the amount of money received from a loan transaction

   b. **interest**—the charge that is paid for use of money borrowed from an institution or person

   c. **interest rate**—the charge paid for money borrowed, expressed as a percentage of the principal

   d. **time**—the length of payback for money borrowed, usually expressed in terms of years

   e. **true** or **actual annual rate**—a standard or basic interest rate for comparison of different methods of charging interest

   f. **simple interest**—the interest rate on a loan with a single (one) payment

   g. **interest on the unpaid balance**—a method of charging interest that causes the interest payments to decline as the principal loan balance declines

   h. **discounted interest**—a type of loan that requires the interest to be paid in advance or at the time the loan is received. The interest is deducted from the original amount of the loan, leaving the borrower with less than the amount actually borrowed

   i. **add-on-interest**—usually charged on installment loans with interest charged on the full amount of the loan for the entire time of the loan

2. Study the following examples:

   Simple Interest

   a. $12,000 is borrowed for one year at 10 percent interest.

   Formula: $I = PRT$

   $$I = \$12{,}000 \times 10\,\% \times 1 \text{ (year)} = \$1{,}200$$

   The simple interest rate is also the true annual rate: 10%.

b. $12,000 is borrowed at 10 percent interest, $6,000 will be repaid at the end of 6 months, and the remaining $6,000 will be paid at the end of the year.

$I = PRT$

$I = \$12{,}000 \times 10\ \% \times \frac{1}{2}\ \text{(year)} = \$600$

$I = \$6{,}000 \times 10\ \% \times \frac{1}{2}\ \text{(year)} = \$300$

$\$900$

When interest is paid only on the unpaid balance, the stated interest rate is the same as the true annual rate: 10 percent.

Discounted Interest

c. $12,000 is borrowed at 10 percent but is discounted so $1,200 interest is deducted.

$$\text{rate} = \frac{\text{Interest}}{\text{Adjusted principal} \times \text{time}} \times 100$$

$$= \frac{\$1{,}200}{(\$12{,}000 - 1{,}200) \times 1}$$

$$= \frac{\$1{,}200}{\$10{,}800 \times 1}$$

$$= 11.11\ \%$$

d. $12,000 is borrowed at 10 percent for 3 years with 36 equal payments.

$I = PRT = \$12{,}000 \times 10\% \times 3\ \text{(years)} = \$3{,}600$

$$\text{true rate} = \frac{2 \times \text{no. of payment} \times \text{interest charged}}{\text{Principal} \times \text{time} \times (\text{no. of payments} + 1)}$$

$$= \frac{2 \times 36 \times 3{,}600}{12{,}000 \times 3 \times 37}$$

$$= \frac{259{,}200}{1{,}332{,}000}$$

$$= 19.46\ \%$$

e. Chart 1 shows the difference in interest paid and true annual interest rate for different types of interest charged on the following.

Amount borrowed = $12,000
Interest rate = 10 %

Kind of Interest	Rate	Time (Years)	Interest Paid	True Annual Interest Rate
Simple Interest	10%	1	$1,200	10%
Interest on Unpaid Balance	10%	1	$ 900	10%
Discounted Interest	10%	1	$1,200	11.11%
Add-on Interest	10%	3	$3,600	19.46%

Observations

1. You are planning to borrow $15,000. You have been offered a 12 percent rate with the following options:

   a. Simple interest for one year.

   b. Interest on unpaid balance for one year with half due in 6 months and half due at the end of the year.

   c. A discounted loan for one year.

   d. Add-on-interest for three years with 36 payments.

   Calculate the amount of interest and true APR for each of the preceding options and record them in the following chart.

Kind of Interest	Rate	Time (Years)	Interest Paid	True Annual Interest Rate
Simple Interest	12%	1		
Interest on Unpaid Balance	12%	1		
Discounted Interest	12%	1		
Add-on Interest	12%	3		

Conclusions

1. What is the lowest interest rate? What is the highest?

2. If you were to make payments each month or put in the bank an amount equal to a payment (for those due in one lump sum), which would cost the most per month? Which would cost the least?

3. Based on the answer to conclusion 2, why do you think some people would choose an add-on-interest loan?

EXERCISE 51: In the Money—Writing a Business Plan

Materials Needed

- ✓ Computer
- ✓ Internet access
- ✓ Calculator

Purpose

Starting a new business is a complicated process and must start with a plan. In this exercise, you will practice writing a simple business plan.

Procedure

In this exercise, you will be creating a business plan for a business of your choice. If you currently are involved in an agribusiness, use that information for your business plan. If not, use your imagination to create one. As you answer the questions below, estimate any dollar amounts you need. The purpose of this exercise is to understand the process involved in the planning, not focusing on actual dollar amounts.

1. Determine what product or service your business will provide.
2. Who is your target customer?
3. What makes your product special and you committed to the business?
4. Estimate what kinds of permits, insurance, location costs, loan payments and other expenses you may have.

Observations

Fill out the estimated expense table below.

Expense	Cost/Month

5. Add your costs.
6. Taking costs into consideration, how much will you charge for the product or service?

Conclusions

Write a one page paper describing your plan.

1. Include an introduction that describes your business.
2. In the body of the paper, describe your target customer and why they would want or need your service or product.
3. Describe what makes your product unique and why you are personally committed to your business.
4. Briefly explain the following; your estimated costs, what you will need to charge for your product or service, and how much you will have to sell to make a profit.
5. Write a summary of the information you provided in the context of the paper. Be sure to include the information that will show your business can be profitable.

EXERCISE 52: IT'S A BALANCING ACT—BALANCE SHEETS

Materials Needed

- ✓ Pencil
- ✓ Calculator

Purpose

To demonstrate the use of balance sheets in agribusiness.

Procedure

1. Study the following terms:

a. **balance sheet**—a form showing the financial condition of a business at a definite point in time. It lists all assets, value of assets, and liabilities of the business. It is also known as a net worth statement, financial statement, or statement of financial condition

b. **assets**—the property or resources owned and controlled by a business

c. **current assets**—cash or other assets that can be converted to cash within 12 months. Examples: checking and/or savings accounts; accounts receivable; inventory held for sale (market animals and stored crops); other near-cash items such as securities, stocks and bonds; and cash value of life insurance

d. **intermediate assets**—resources or production items with useful lifetimes of 1 to 10 years. Examples: equipment, machinery, and breeding stock. These assets are generally depreciable

e. **fixed assets**—also called long-term assets; include permanent items such as real estate and improvements on buildings. The useful lifetime is generally over 10 years

f. **liabilities**—all the debt obligations of the business

g. **current liabilities**—debts due within the operating year, that is, a 12-month period. Examples: notes and accounts payable, rents, taxes, interest, and principal payments due on intermediate or long-term debt within the next 12 months

h. **intermediate liabilities**—nonreal-estate debt that corresponds to intermediate assets. Loan terms are normally for a period of 1 to 10 years. Examples: improvements to real estate, equipment purchases, breeding livestock and dairy stock, and capital requirements for major adjustments in farm operation

i. **long-term liabilities**—mortgages and land contracts on real estate minus principal due within 12 months

j. **net worth**—total assets minus total liabilities; also called owner's equity

k. **liquidity**—the ability of a business to make enough cash to pay bills without disrupting business. One of the best measures of liquidity is current ratio.

l. **current ratio** $= \dfrac{\text{Current assets}}{\text{Current liabilities}}$

m. **intermediate ratio**—the intermediate liquidity of the business can be determined by the intermediate ratio.

$$\text{intermediate ratio} = \frac{\text{Current} + \text{intermediate assets}}{\text{Current} + \text{intermediate liabilities}}$$

n. **solvency**—the ability to pay debts; shows the amount that would be left after all assets are converted to cash and all debts paid. Determined by:

$$\text{net capital ratio} = \frac{\text{Total asset}}{\text{Total liabilities}}$$

o. **equity**—the owner's share of the business. Shown by debt-equity ratio; also known as the debt-to-net worth ratio.

$$\text{(debt-to-net worth ratio)} = \frac{\text{Total liabilities}}{\text{Net worth (owner's equity)}}$$

Lenders are interested in this ratio because they usually prefer to provide loans that are equal to less than 50 percent of assets. Lenders prefer a debt-equity ratio of less than 1.

2. Using the following information,

Current assets = $25,000	Current liabilities = $15,000
Intermediate assets = $75,000	Intermediate liabilities = $50,000
Fixed assets = $200,000	Long-term liabilities = $150,000

a. Figure the current ratio.

b. Determine the intermediate ratio.

c. Use the net capital formula to determine the net capital ratio.

d. Determine the net worth.

e. Determine the debt-equity ratio.

Observations

1. Use the following information to complete the accompanying balance sheet.

Value of machinery	$75,000.00
Loan on feeder steers	$18,000.00
Value of cow herd	$21,000.00
Mortgage on land	$125,000.00
Checking account balance	$2,700.00
Value of land	$337,000.00
Bank note due next month	$16,500.00
Balance on machinery loan	$34,000.00
Value of feeder steers	$29,000.00
Value of stored grain	$23,00.00

a. Fill in the current assets.

b. Fill in the current liabilities.

c. Complete the intermediate assets.

d. Complete the intermediate liabilities.

e. What are the fixed assets? List them on the balance sheet.

f. Add the long-term liabilities to the sheet.

g. Determine the net worth.

BALANCE SHEET

Current Assets	Current Liabilities
__________	__________
__________	__________
__________	__________
Intermediate Assets	**Intermediate Liabilities**
__________	__________
__________	__________
Fixed Assets	**Long-Term Liabilities**
__________	__________
Total Assets __________	**Total Liabilities** __________
Net Worth __________	

Conclusions

1. What is the current ratio?

2. What is the intermediate ratio?

3. What is the net capital ratio?

4. What is the debt-equity ratio?

5. Is the business solvent?

EXERCISE 53: Rich Man, Poor Man–Profit–Loss Statements

Materials Needed

✓ Pencil

✓ Paper

✓ Calculator

Purpose

To demonstrate the use of profit–loss statements in agribusiness.

Procedure

1. Study the following terms:
   a. **profit and loss (income) statement**—the financial record that reflects the profitability of the business over a specified period of time
   b. **income**—cash income includes money received from the sale of crops, livestock, and livestock products during the year. It also includes custom work and government payments. Gain from the sale of capital assets is also entered as cash income; the entry is the difference between the selling price and the book value. Non-cash income includes products used in the home or given to a farm for custom work. Positive changes in inventory are also considered non-cash income.
   c. **expenses**—cash expenses are monies paid to operate the business, such as those for seed, feed, fertilizer, market livestock, and fuel. Fixed-cash expenses include taxes; interest; insurance; and, in some cases, repairs. Non-cash expenses include depreciation on machinery, equipment, buildings, and purchased breeding stock. Negative changes in inventory are recorded as non-cash expenses. Not recorded as expenses are costs of any new capital assets and principal payments on loans.
   d. **cash accounting method**—income and expenses are recorded only when actual cash transactions occur. Most non-cash income and expenses are not recorded under the cash method. Depreciation is an exception, as is the value of farm products used in the home. Both of these are included under the cash method.
   e. **accrual method**—records income when it is earned and expenses when they are incurred. Inventory changes are included in the accrual method. Because of this, it is actually a more accurate method of accounting
   f. **net cash income**—the difference between total cash income and total cash expenses

g. **net farm income**—net cash income plus non-cash adjustments

h. **return to capital** or **return on investment**—a measure of profitability based on a ratio obtained by dividing the return to total capital by total farm assets

i. **return to capital**—adjusted net farm income minus opportunity cost of labor minus opportunity cost of management

j. **ratio of return to capital (%)** $= \dfrac{\text{Return to total capital}}{\text{Total farm assets}} \times 100$

k. **adjusted net farm income**—net farm income plus interest paid minus value of unpaid family labor

l. **return to labor and management**—that portion of net farm income that remains to pay the owner for personal labor and management after capital is paid a return equal to its opportunity cost. Return to labor and management equals adjusted net farm income minus opportunity cost on total capital.

m. **return to labor**—return to labor and management minus opportunity cost of management

n. **return to management**—return to labor and management minus opportunity cost of labor

o. **return to equity**—net farm income minus opportunity cost of own labor minus opportunity cost of management minus value of unpaid family labor

p. **rate of return to equity (%)** $= \dfrac{\text{Return to equity}}{\text{Net worth}} \times 100$

2. Working with formulas:

Net farm income	$ 80,000.00
Value of operator's management	$ 5,000.00
Owner's equity	$500,000.00
Interest paid	$ 25,000.00
Value of total assets	$750,000.00
Value of operator's labor	$ 15,000.00
Opportunity cost of capital	10%

a. What is the return to capital in percent?

b. What is the return to management?

c. What is the return to equity in percent?

Observations

1. Use the following information to complete the profit and loss statement:

Grain sorghum sales	$30,000.00
Alfalfa hay sales	$9,500.00
Building depreciation	$2,800.00
Increase in inventory of livestock	$15,000.00
Feeder calf sales	$41,000.00
Total farm cash operating expenses	$57,000.00
Machinery depreciation	$12,000.00
Decrease in crop inventory	$4,000.00

Conclusions

1. What is the total revenue?

2. What are the total expenses?

3. What is the net farm income?

4. What efficiency factors can be determined using the information from a profit–loss statement?

EXERCISE 54: Money, Money, Money—Interest

Materials Needed

✓ Paper

✓ Pencil

✓ Calculator

Purpose

Interest is calculated in different ways in the business world. The purpose of this lab is to learn how different types of interest are calculated.

Procedure

1. The formula for calculating basic simple interest is as follows;

   $I = PRT$

   $I =$ Total interest

   $P =$ Principal (loan amount)

   $R =$ Interest rate as a decimal

   $T =$ time

   Example:

   A loan for \$1,000 is taken at 10 percent interest to be paid at the end of 5 years. What is the total amount of interest to be paid?

   $I = (1{,}000)(.10)(5)$

   $I = \$500$

2. The formula to calculate the future value of money with simple compound interest is as follows:

   $\text{FV} = \text{PV} \times (1 + r)^n$

   FV = Future value of money

   PV = Present Value (at start)

   $r =$ interest rate as a decimal

   $n =$ number of periods (usually years)

   Example:

   You invest \$500 in a money market at a 6 percent interest rate. What will the future value be in 5 years?

   $\text{FV} = 500 \times (1 + .06)^5$

   $\text{FV} = \$669.11$

3. The formula to calculate compound interest is as follows:

$$A = P\left(1 + \frac{r}{n}\right)^{nt}$$

P = principal amount (initial investment)

r = annual interest rate (as a decimal)

n = number of times the interest is compounded per year

t = number of years

A = amount after time (t)

Example:

A loan is taken out for 10,000 with an interest rate of 3 percent that is compounded quarterly (four times a year). How much interest will be paid at the end of the loan in 5 years?

$A = 10{,}000(1 + .03/4)^{4\times 5}$

$A = 10{,}000(1.0075)^{20}$

$A = 10{,}000(1.161184)$

$A = 11{,}611.84$

$10{,}000 - 11{,}611.84 = \$1{,}611.84$ total interest paid

Observations

1. Jake took out a basic simple interest loan for $1,000.00 at a 6 percent. Using the formula given in procedure #1, figure how much interest he will pay by the time the loan is paid off in five years.
2. Using the simple compound interest formula from procedure #2, calculate the future value (total amount he will need to repay) on Jake's loan if he was given a simple compound loan instead of a basic simple interest loan.
3. Using the compound interest formula given in procedure #3, figure the amount of interest Jake would have to pay if he took out a compound loan.

Conclusions

1. How much more interest would Jake have paid if he were to have had a compound interest loan instead of the basic interest loan?
2. When considering barrowing money, why is it important to understand what type of interest will be applied to a loan?

EXERCISE 55: Can't We All Just Get Along—Cooperative Math Simulations

Materials Needed

- ✓ Pencil
- ✓ Paper
- ✓ Calculator
- ✓ Textbook

Purpose

In this exercise, you will use some basic math skills that are used in cooperative businesses.

1. In the United States, 2,000 marketing cooperatives serve 24,000,000 farmers for a total volume of $10 million. On the average, what dollar amount did each cooperative contribute to the total volume?

2. Paul Stark did $2,000 worth of business with his supply cooperative last year. At the end of the fiscal year, Paul received a $70 patronage refund. What percent of his purchases did the refund represent?

3. You are a member of a marketing cooperative with a membership of 920. The cooperative by-laws provide one vote per membership plus an additional vote for each 10 tons of fruit marketed. Assuming that all members voted, 2,500 votes were tallied, and you marketed 5 percent of the total tonnage marketed, how many votes were you allowed to cast?

4. An egg marketing cooperative withholds $.02 per dozen for its revolving fund. A member is credited with $2,345 in the fund during 2002. If his production average was 175 eggs per hen, how many laying hens did he have in his flock?

5. An egg marketing co-op sells its eggs in cases of 30 dozen. If the co-op withheld $.01 cent per dozen for the revolving fund and it credited a member with $2,200, how many 30-dozen cases did the member market in that year?

6. James Dillon did $7,500 in business with his supply co-op last year. At the end of the year, he received a patronage refund in the amount of $316. What percent of his year's purchases did this cash refund represent?

7. A member of a fruit-marketing co-op delivered 900 tons of fruit to his co-op. Sixty-five percent was sold as fresh fruit, netting $150 per ton; 30 percent was sold for canning, netting $125 per ton; and 5 percent was sold for drying, netting $85 per ton. If the co-op retains 4 percent of all sales for its revolving fund, what did the member receive in cash for the season?

Conclusions

1. What are the benefits of a cooperative business?
2. If you were interested in starting a co-op business, how would you go about recruiting others to join?